PERGAMON INTERNATIONAL LIBRARY
of Science, Technology, Engineering and Social Studies

*The 1000-volume original paperback library in aid of education,
industrial training and the enjoyment of leisure*

Publisher: Robert Maxwell, M.C.

D1740072

Basic Electric Circuits

Other Titles of Interest in the Pergamon International Library

Basic Electric Circuits

SECOND EDITION

by

A. M. P. BROOKES, M.A., A.M.I.MECH.E.

*Fellow of St. John's College, Cambridge
and Lecturer in Engineering
at the University of Cambridge*

PERGAMON PRESS

OXFORD · NEW YORK · TORONTO
SYDNEY · PARIS · BRAUNSCHWEIG

U.K.   Pergamon Press Ltd., Headington Hill Hall, Oxford OX3 0BW,
                      England

U.S.A.   Pergamon Press Inc., Maxwell House, Fairview Park,
                      Elmsford, New York 10523, U.S.A.

CANADA   Pergamon of Canada Ltd., 207 Queen's Quay West,
                      Toronto 1, Canada

AUSTRALIA   Pergamon Press (Aust.) Pty. Ltd., 19a Boundary Street,
                      Rushcutters Bay, N.S.W. 2011, Australia

FRANCE   Pergamon Press SARL, 24 rue des Ecoles,
                      75240 Paris, Cedex 05, France

WEST GERMANY   Pergamon Press GmbH, 3300 Braunschweig, Postfach 2923,
                      Burgplatz 1, West Germany

Copyright © 1975 Pergamon Press Ltd.

First edition 1966

Second edition 1975

Library of Congress Cataloging in Publication Data

Brookes, Alexis Michael Panther.
Basic electric circuits.

Includes index.
1. Electric circuits.   1. Title.
TK454.B76 1975   621.319'2   75-8774
ISBN 0-08-018310-7
ISBN 0-08-018309-3 pbk.

Printed in Great Britain by Biddles Ltd., Guildford, Surrey

Contents

Preface

THERE are many excellent books on advanced circuit analysis, but singularly few on the elements of this branch of electrical engineering, the topic usually being treated in one or two chapters in comprehensive electrical textbooks.

Familiarity with the elements of electric-circuit theory contributes greatly towards the full understanding of many branches of electrical engineering, both light and heavy and, by the use of the technique of equivalent circuits, a very large range of problems may be transformed into circuits capable of quick and straightforward solution.

To a large extent this book is based on lectures given to undergraduates in their first or second year of a University engineering course, and it is hoped that it will produce a solid grounding in the elements of circuit analysis both for university students and for those students reading engineering at technical colleges and other similar institutions. The earlier chapters should not be beyond the scope of sixth-form secondary school students.

Typical examination questions have been given at the end of each chapter and there are a number of worked examples in the text. The examples and questions have been widely chosen so as to illustrate various applications of the ideas discussed in any particular chapter, and duplication of an example on any specific aspect has been avoided as far as possible as it is felt that by far the best source of examples for practice is a collection of past examination papers for the examination for which any particular student is preparing. In general the M.K.S. system of units has been used.

The physical appearance and construction of common circuit elements has been included, together with some mention of stability and accuracy, since experience shows that many students, when they come to do their practical work, have very little idea of the size, shape or suitability of the various components used in their experiments. An engineer is essentially a practical man and the sooner he gets the " feel " of his components and apparatus the better.

During the few years since the original publication of *Basic Electric Circuits* considerable advances have been made in the semiconductor field, particularly in respect of integrated circuit techniques. This second edition of the book includes considerable new material on transistors and semiconductor devices, and its general coverage has been considerably widened.

Cambridge   A. M. P. BROOKES

Acknowledgement

The tables given on pp. 30-33 are printed by the kind permission of Litton Educational Publishing Inc., New York, U.S.A.

CHAPTER 1

Conducting and Insulating Materials

IF A MATERIAL will allow an electric current to flow through it easily it is known as a *conductor* and if it effectively prevents the flow of current it is known as an *insulator*. The majority of common materials fall definitely into one class or the other and the difference between the two classes is very considerable. If for instance two rods of the same size were made, one from a typical conducting material and the other from a typical insulating one, the ratio of the ability to conduct electricity would be of the order of 10^{20}. This enormous difference is due to the way in which an electric current is carried through a material. All materials are made up of atoms which themselves consist of a dense central part called a nucleus surrounded by a cloud of electrons. The nucleus has a positive charge and the electrons have a negative charge. The electrons are held in the atom by electric forces but the number and arrangement of electrons is different for different elements with the result that certain materials have electrons very closely held by these forces. In others some of the electrons on the surface of the atoms can be fairly readily detached so that they can wander about in the material from atom to atom. The closely bound electron system is a characteristic of insulators and the free outer electron system is characteristic of conductors.

Figure 1.1 shows in simplified form the arrangement of electrons around different nuclei. It will be seen that the nucleus can contain one or several positive particles or protons and similarly that the number of electrons can be one or more. There are, however, always the *same* number of protons as electrons in the atom in its normal state because each proton carries one positive charge and each electron one negative charge and these must balance if the atom as a whole is to be neutral. It can also be seen that the electrons are arranged in circles in the diagram which correspond to spherical shells designating the orbits of the electrons. There is a maximum number of electrons which will remain in any one shell and hence when one shell is full another at a greater radius is started. The number of protons and electrons that an atom contains differs for every element and, due to the shell arrangement and the finite number of electrons which each shell can contain, it will be clear that some elements will have outermost

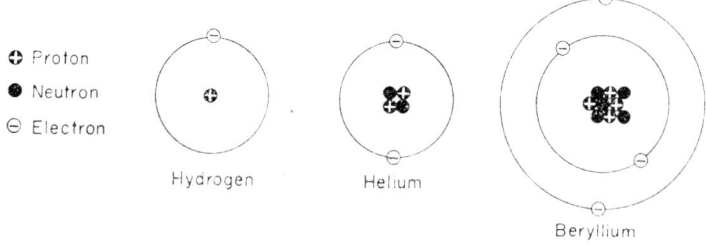

Fig. 1.1

shells which are full whilst others will have several places unoccupied. It is this difference which accounts for the ease or otherwise with which an electric current can pass through a material because a material with its outer shell full does not easily lose or take up an electron whilst, if there is an unoccupied place in an outer shell, an electron can slip into it and another electron leave the shell much more readily.

In metals the outer shells have not a full complement of electrons and also the atoms are closely packed together so that electrons can wander from atom to atom readily. In the absence of any unidirectional attractions the electrons will wander at random and on the average no net charge will be carried from one part of the metal to another as shown in Fig. 1.2.

However if at one point in the metal, at the end of a wire for instance, electrons are being attracted away by any means then the motion of electrons in the vicinity will no longer be completely random and there will be a drift towards the place from which they are being removed, as shown in Fig. 1.3, this drift constituting an electric current.

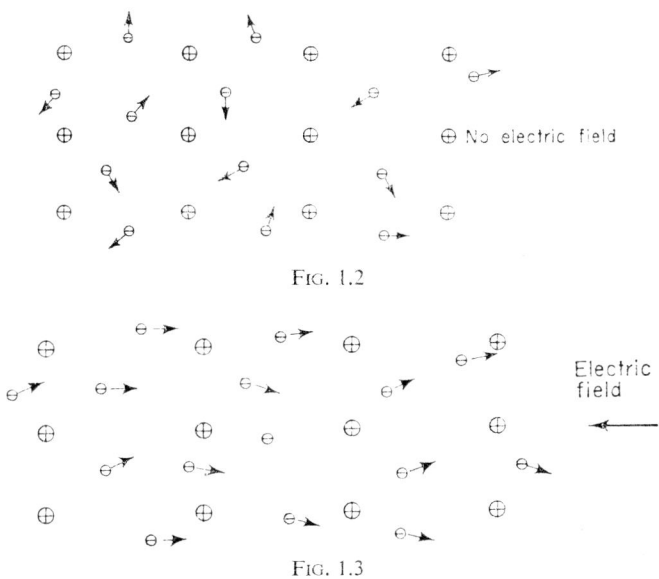

FIG. 1.2

FIG. 1.3

It is important to distinguish between the velocity of this drift and the velocity of propagation of the change from random motion. If an electron drift is established at one end of a wire this drift will propagate through the wire to the other end in a time very much shorter than the time it would take an electron travelling at the mean drift speed to travel the same distance. This is analogous to a transverse vibration being transmitted down a taut string where the transverse velocity of the parts of the string is quite unconnected with the rate at which the disturbance is propagated along the string.

Not only is the drift velocity of the electron small compared with the velocity of propagation, it is also much smaller than the random motion velocity. Suppose that a conductor has a cross-sectional area A and that there are n free electrons, of charge e, per cubic metre moving with drift velocity v. The current I flowing will be given by

$$I = A.n.e.v. \text{ amps}$$

$$\therefore \quad v = \frac{I}{A.n.e}$$

For a copper wire of square section of side 1 mm and a current of 1 amp we have:

$$A = (10^{-3})^2 = 10^{-6}$$
$$I = 1$$
$$n = 8\cdot3 \times 10^{28}$$
$$e = 1\cdot6 \times 10^{-19}$$

which gives $v = 7\cdot5 \times 10^{-5}$ m/sec.

As would be expected from the above description of the mechanism of conduction the ability to conduct an electric current is not the same for all conducting materials nor is it the same at all temperatures. Changes in temperature imply changes in the energy of the electrons in the atoms of a substance and, since the mobility of these electrons governs the ease with which the electricity can be conveyed, it is natural that temperature and conduction should be interrelated. The property of opposing conduction is known as the *resistivity* of a material, symbol ρ, and the reciprocal $1/\rho$ is known as the *conductance*. It is more usual to specify materials by their resistivities and to give these a unit *ohm-metres* or *ohm-centimetres*. Numerically the resistivity of a homogeneous material at a given temperature could be obtained by the measurement of the current I through a cube of edge one metre when a potential difference V volts was applied to one pair of opposed faces of the cube. If the experiment were performed on a cube of edge one centimetre the result would, of course, then be in ohm-centimetres.

Table 1 gives the resistivities of several common conducting materials in order of ability to conduct.

The last three items in the table are alloys especially developed to have fairly high values for their resistivities so that resistors may be manufactured without requiring large quantities of wire. Brass is also an alloy and has different values of resistivity for different proportions of its constituent elements.

TABLE 1

Resistivities of Metals at 20°C

	ohm-metre
Silver	1.62×10^{-8}
Copper	1.76×10^{-8}
Gold	2.40×10^{-8}
Aluminium	2.83×10^{-8}
Tungsten	5.48×10^{-8}
Nickel	7.24×10^{-8}
Brass	8×10^{-8}
Iron	9.4×10^{-8}
Platinum	10×10^{-8}
Lead	20×10^{-8}
Manganin	45×10^{-8}
Eureka	49×10^{-8}
Nichrome	108×10^{-8}

Since the variation of resistivity with change in temperature can be very large for some materials and is quite considerable for many, it is as necessary to have a method of estimating this change as it is to know the resistivity itself. The relation between resistivity and temperature is not a linear one as can be seen from Fig. 1.4 which is for lead. It will also be noted that there is a sudden change at the temperature at which lead melts. Thus any accurate figure for resistivity at a given temperature is best taken from an experimentally obtained curve for the material. However, in many practical circumstances the variation in temperature from one at which the resistivity is known is not too great and a linear relation can be used. This relation is usually of the form

$$R_T = R_1[1 + \alpha_1(T - T_1)]$$

where R_T is required resistivity at temperature T

R_1 is the known resistivity at temperature T_1

α_1 is the *temperature coefficient of resistance* for the material appropriate to T_1.

It is important to note that α_1 will be different for different values of T_1 because the assumption we are making in assuming a linear relation is that we may extrapolate linearly from any point on

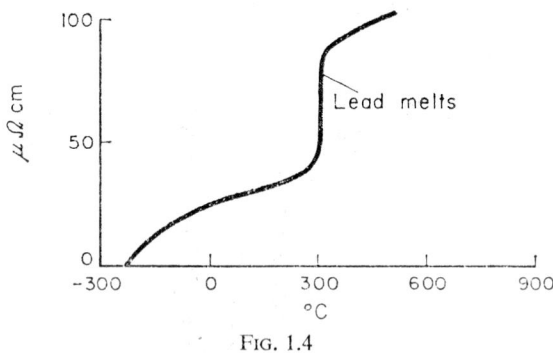

FIG. 1.4

our actual true curve for the material. This means that at any given point corresponding to T_1 on the curve we must draw a tangent and work along it. Obviously the slope of the tangent will vary with the value of T_1 and hence the value of α_1 will vary.

It is not usual, however, to give extensive tables of α for large numbers of temperature values and so Table 2 gives values of α_0 which assumes the resistivity T_0 is known at 0°C.

It will be noticed that Eureka may have a small *negative* temperature coefficient. Carbon is another substance which has a negative coefficient which is about -0.0005. Some other materials consisting for example of oxides of manganese, nickel, iron, zinc, titanium and magnesium have very large negative temperature coefficients and also have high resistivities compared with the pure metals and alloys we have been considering, although not nearly so high as for insulating materials. These materials are known as *semi-conductors* and are extremely important as the

TABLE 2

Temperature Coefficients of Resistance

	°C units
Manganin	0·000002
Eureka	− 0·00007 to +0·00004
Nichrome	0·0001
Brass	0·001
Gold	0·0034
Platinum	0·0037
Silver	0·0038
Copper	0·0043
Lead	0·0043
Aluminium	0·0043
Tungsten	0·0045
Iron	0·0055
Nickel	0·0059

essential part of transistors, metal rectifiers, thermistors and various other devices. As an example the graph in Fig. 1.5 shows a typical thermistor temperature–resistivity characteristic where the resistivity is plotted to a log scale and indicating a change in resistivity of more than a million to one for a temperature change of from −100°C to +400°C. Because of this very large change in resistance with temperature it is simple to use a thermistor as a very sensitive thermometer since a relatively rough measurement of its resistance will be sufficient to determine its temperature to within ±0·01°C once its calibration is known.

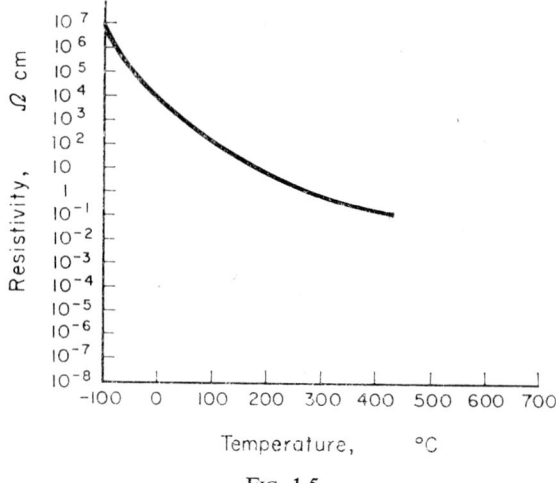

FIG. 1.5

The two most used semiconductor materials are germanium, with a resistivity at room temperature of about 47 ohm-cm, and silicon of resistivity 3×10^5 ohm-cm under the same conditions. Other important materials in use in various devices are gallium arsenide, lead sulphide, indium antimonide and other compounds each of which has some special useful property, but silicon devices have now become the commonest.

To understand the working of a semiconductor device it is first necessary to consider the crystal structure of the basic material. Fig. 1.6 shows in diagram form the arrangement of the nucleus and electrons in a silicon atom, from which it can be seen that the outer shell of electrons has four members known as the valence electrons. The full permitted complement for this shell is eight electrons so that when the atoms are assembled regularly in a crystal of the material it is found, upon examination of the crystal structure, that each of the valence electrons of any one atom is shared with the four neighbouring atoms so that effectively every atom has a *full* shell, as is shown diagrammatically in Fig. 1.7. When of the structure shown in Fig. 1.7 silicon has a resistivity of 3×10^5 ohm-cm at room temperature, as stated above, when the material is said to be *intrinsic*, the effect of thermal agitation enabling a few electrons to break away from the parent atoms and become current carriers. However, by the addition of an extremely small amount of a pentavalent impurity, which thus has five valence electrons instead of the four for pure silicon, the resistivity can be reduced to only a few ohm-cm, the extra valence electron being very free to move about the lattice. Only one impurity atom for every 10^8 atoms of the pure silicon is necessary to produce this very large change in

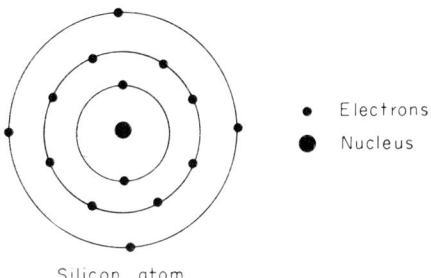

• Electrons

● Nucleus

Silicon atom

FIG. 1.6

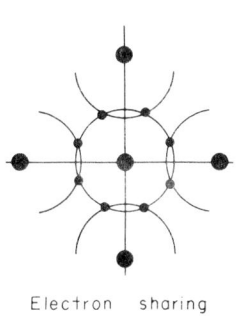

Electron   sharing

in   Germanium

or   Silicon

FIG. 1.7

resistivity and the impurity atom is known as a *donor* atom producing, as it does, an extra electron in the lattice, making the material into n-type. This is shown diagrammatically in Fig. 1.8.

It is also possible to "dope" pure silicon with a few trivalent or *acceptor* atoms having only three electrons in their outer shells. In this case, as shown in Fig. 1.9, there will be a vacant space or "hole" in the lattice electron system and the subsequent material is known as p-type because the lack of an electron is equivalent to a positive charge and the vacant space can be considered to move about because any free electrons in the vicinity which have come from neighbouring atoms can occupy holes and thus effectively transfer these holes to the atoms from which the electrons have come originally. For convenience it is usual to consider n-type material as having a number of free negative charges available to carry current by moving through the lattice and p-type material as having a number of positive charges available to carry current also by moving through the lattice, and this system is shown in Fig. 1.10.

It must be emphasized that the impurity atoms, both of donor and acceptor type, are fixed rigidly in the lattice and will have a charge equal in magnitude to that of one electron, this charge being positive for a donor atom—because it has effectively lost one of its five electrons—and negative for an acceptor atom—because it has accepted an extra electron above its original three. Hence, if by some means the mobile charges can be swept away from some part of the doped material, a space charge will exist in that region due to the charges on the doping atoms.

Under room temperature conditions there will always be a few electrons thermally removed from their parent atoms and thus a few holes simultaneously generated. These charges are known as *minority* carriers to distinguish them from the carriers introduced by doping which are therefore called *majority* carriers as indicated in Fig. 1.10. Thermal generation and recombination can be considered as taking place continuously at all temperatures above absolute zero and this is the main reason for sensitivity to temperature shown by all semiconductor materials.

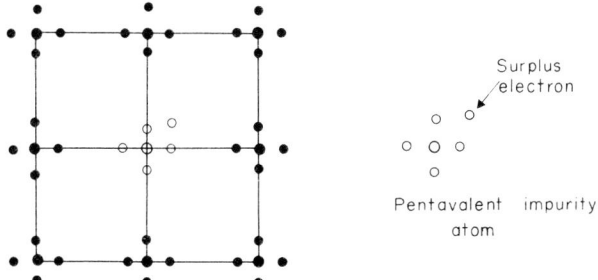

n – type semiconductor material

FIG. 1.8

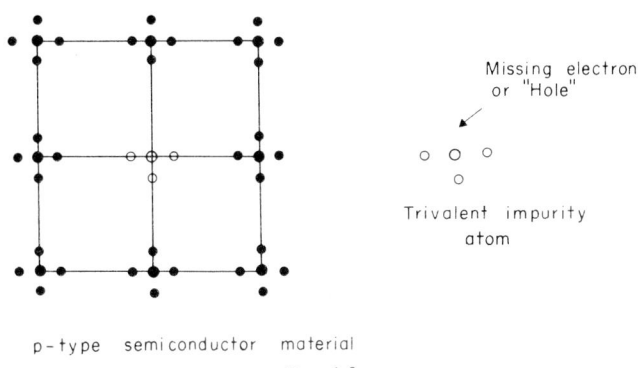

p – type semiconductor material

FIG. 1.9

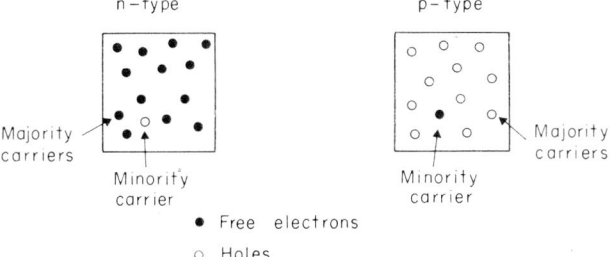

FIG. 1.10

At very low temperatures between $-268°C$ and $-273.15°C$, which is absolute zero, many metals have an abrupt change in their temperature–resistivity characteristic, there being a sudden fall in resistivity at a particular temperature with the resultant value being so low in some instances that the metal may be said to have negligible resistance to the passage of a current. This phenomenon is called *superconductivity* and nineteen elements exhibit it including common ones such as aluminium, tin, mercury and lead.

An electric current can be conducted readily through liquids and gases and by streams of electrons in a vacuum but these types of conduction will not be considered here.

We shall next consider the materials which have such high values of resistivity that they are normally used for preventing the passage of a current rather than facilitating it. These materials are known as *insulators* or *dielectrics* and are characterized by their atoms having the maximum permissible number of electrons in their outer shells. There is thus no mobility of electrons and if the material is put in an electric field the orbits of the electrons are distorted as shown in Fig. 1.11, but there is no redistribution of electrons, as in a conductor, so as to make the field within the material zero. There is thus a field *inside* the body of an insulator placed in an electric field. When the surface atoms of the insulator are distorted as shown in Fig. 1.11 the dielectric material is said to be *polarized* and an atom of the material is then called an induced dipole. Some of the strength of the applied field is

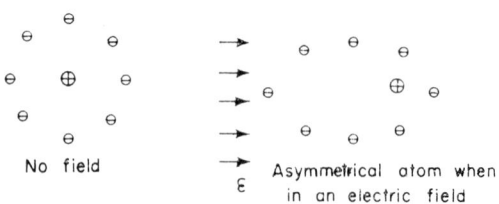

No field

Asymmetrical atom when in an electric field

FIG. 1.11

absorbed in making the induced dipoles so that the net field, after passing inside the insulator, is less than at entry. This field-reducing property is a measure of the effectiveness of the insulating material and is different for different substances. It is reasonably constant for any one material and is called the dielectric constant, κ, or alternatively ε_r.

Let us now consider the effect of sandwiching a dielectric material between two conducting plates as shown in Fig. 1.12.

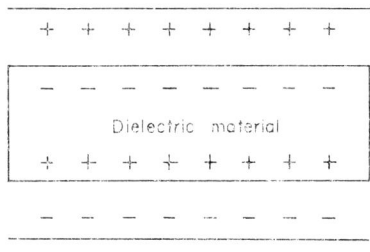

FIG. 1.12

The plates may each be assumed to be *one* metre square and to carry a charge, Q coulombs. Now in accordance with the above discussion there will be induced dipoles on the surface of the dielectric of total charge q say. From experiment it has been found that q is linearly related to the electric field $\mathscr{E}$ inducing it and the constant of proportionality is known as the electric susceptibility χ.

Thus for our square metre

$$q = \chi \mathscr{E}$$

Now taking both Q and q into account and noticing that where they appear adjacent they are opposite in sign, we find that the net charge Q' at the opposite faces of the dielectric is given by

$$Q' = Q - q$$

But the resulting electric field must be given by the equation

$$\mathscr{E} = \frac{Q'}{\varepsilon_0} \text{ volts/m}$$

$$= \frac{Q - q}{\varepsilon_0}$$

$$= \frac{Q - \chi\mathscr{E}}{\varepsilon_0}$$

$$\therefore \mathscr{E} = \frac{Q}{\varepsilon_0(1 + \chi/\varepsilon_0)}$$

It is the dimensionless quantity in brackets $(1 + \chi/\varepsilon_0)$, which is the constant κ referred to above.

We can thus finally obtain the following expressions:

$$\mathscr{E} = \frac{Q}{\kappa\varepsilon_0}$$

$$V = \mathscr{E}d$$

$$= \frac{Qd}{\kappa\varepsilon_0}$$

The charge in a capacitor is given by:

$$Q = CV$$

$$\therefore \quad C = \frac{Q}{V} \text{ farads} = \frac{\kappa\varepsilon_0}{d}$$

It is clear from this result that if there had been no dielectric inserted between the plates the capacity would have been given by

$$C_0 = \frac{\varepsilon_0}{d} \text{ farads}$$

Thus the insertion of a dielectric has increased the capacitance by the factor κ. The product $\kappa\varepsilon_0$ is known as the permittivity of the dielectric.

Nothing has yet been said concerning any limit to the potential gradient $\mathscr{E}$ but this is of considerable practical importance because if it is exceeded the dielectric will in most cases be permanently damaged. When $\mathscr{E}$ is too great electrons are forcibly removed through a part of the material and in those parts a current of considerable density will flow. This critical value of $\mathscr{E}$ is known as the *dielectric strength* of the material and is not connected in any way with the dielectric constant.

TABLE 3

	Dielectric constant κ	Permittivity farads/metre $\times 10^{-12}$	Resistivity ohm-metres	Dielectric strength volts/metre
Bakelite	5	44	10^9	10^7
Ceramics	7 to 1000	62 to 8800	10^{12}	10^7
Glass	8	75	10^{12}	3×10^7
Mica	6·5	56	10^{14}	10^8
Oil	3	26	10^{13}	10^7
Paraffin wax	2·1	19	10^{15}	5×10^7
Polythene	2·3	21	10^{15}	10^8
Polystyrene	2·6	23	10^{16}	10^8
Rubber	2·2	19	10^9	5×10^7

It can thus be appreciated that whilst with conducting materials we are mainly interested in the resistivity of the material, with insulators we are often concerned with more than one property at the same time. Table 3 therefore gives values of κ, ε, ρ and dielectric strength for several common insulating materials. It will be seen that no one material has a monopoly of advantageous properties and for this reason the materials are listed alphabetically.

CHAPTER 2

Semiconductor Junction Devices

For a thorough understanding of the behaviour of semiconductor junction devices, the starting point must be a knowledge of the mechanism of electrical conduction in semiconductors.

As has been pointed out in the previous chapter germanium and silicon having less than about one part in 10^{10} impurity content are known as intrinsic semiconductors, because their properties do not alter significantly if they are further purified. The elements are in group IV of the periodic table, and each atom has four valence electrons. The elements form crystals having the same structure as diamond, in which each atom has four nearest neighbours and shares two valence electrons with each, so forming covalent interatomic bonds.

The energy required to release one electron from a bond is about 0·7 eV in germanium and 1·1 eV in silicon, so that at room temperature, a significant number of electrons will be liberated by thermal agitation. Thus in an intrinsic semiconductor, there will exist a number n_i per unit volume of electrons available to take part in electrical conduction. This number will be a function of temperature, and of the semiconductor material.

When an electron is thermally released from a bond in this way, it leaves behind a hole which appears to be positively charged. The liberated electron travels through the lattice until it finds a hole, which it fills by completing the partial bond; this process, electrically equivalent to the combination of a positive and a negative charge, is called recombination. Evidently the rate of recombination will be proportional to the product of the concentrations of holes and electrons, in this case the square of n_i. Since the crystal is in thermal equilibrium, it follows that the rate of generation of hole-electron pairs is also proportional to the square of n_i.

In the absence of an applied electric field, the motion of an electron through the lattice is random. In the presence of a field, the random motion has superimposed upon it a constant velocity in the opposite direction to that of the field. The velocity rather than the acceleration is constant, because the electron repeatedly collides with atoms in the lattice, and loses energy which it has gained by being accelerated since the last collision. The current carried by the electrons is proportional to the field and to the electron density n_i.

There is, however, another effect which contributes to the conduction process. Quantum mechanics predicts that not only thermally liberated, but also bound, electrons may hop into holes. This is equivalent to holes moving about. Analysis shows that if we treat the holes as positively charged particles, then classical mechanics gives the correct results for conductivity.

The importance and usefulness of semiconductors is entirely due to the fact that their properties at normal temperatures are profoundly affected by the presence of minute concentrations of impurities. Suppose that a small quantity of an element from group V of the periodic table is added to the molten semiconductor, and a crystal then grown. The crystal will still have the diamond structure, but each impurity atom will have one valence electron which is surplus to the requirements of the lattice. The energy required to liberate this electron from the parent atom is of the order of 0·1 eV only, so that at normal temperatures all such electrons are free and available to the conduction process: group V atoms are therefore known as donors.

If the number of donor atoms per unit volume is n_d, we now have this concentration of extra free electrons, and the same concentration of extra holes, which this time are fixed, and not available for conduction. If the total concentrations of free electrons and free holes are now n_o and p_o respectively, we have

$$n_o - p_o = n_d \qquad (1)$$

Now the same rate of production of holes and electrons due to thermal agitation will exist as in the intrinsic semiconductor, so that the same rate of recombination must also exist at equilibrium. The rate of recombination is now proportional to $n_o p_o$, so we have

$$n_o p_o = n_i^2 \qquad (2)$$

In the temperature range over which the semiconductor is useful, it will always be the case that n_d is very much greater than n_i in which case, solution of equations 1 and 2 gives

$$n_o = n_d \qquad (3)$$

and
$$p_o = n_i^2/n_d \qquad (4)$$

Thus the number of electrons is vastly greater than the number of holes, and electrons contribute almost all the conductivity. This material is known as n type.

Precisely similar effects occur if the intrinsic semiconductor is doped with an element from group III of the periodic table. In this case the impurity atom has only three valence electrons, and it can accept an electron from elsewhere to complete its fourth bond: it is accordingly called an acceptor, and contributes a hole to the conduction process. If the concentration of acceptors is n_a, we find

$$p_o - n_o = n_a \qquad (5)$$
$$p_o = n_a \qquad (6)$$
$$n_o = n_i^2/n_a \qquad (7)$$

In this case, the number of holes greatly exceeds the number of electrons, and the former are responsible for almost all the conductivity of the material, which is accordingly known as p type material.

In either type of material we see that one polarity of carrier dominates the conduction process: whichever this is, it is called the majority carrier, the other one being the minority carrier. We shall see that the properties of pn junctions depend upon the behaviour of the minority carriers.

In constructing semiconductor devices, cases continually arise in which a material contains both donor and acceptor impurities simultaneously. Such a material has properties determined by whichever impurity is in the majority, and is known as a compensated semiconductor.

The pn Junction

There are several ways in which a pn junction may be produced, and amongst the properties which are affected by the method of production is the abruptness with which the material changes from p type to n type at the junction.

The grown junction has the most abrupt change of type, and is produced as follows. Molten intrinsic material is doped to the required concentration with acceptor dopant, and the growing of a crystal commenced. After a time, donor dopant is added so that from that time onwards, a compensated semiconductor is produced, being n type. The result is a very abrupt change of type at the junction, whose position is well defined. We shall assume such an abrupt junction in what follows, and it will be apparent that only minor modifications are required to the treatment if the junction be less abrupt.

The degree of doping both p and n type regions is such that the concentration of minority carriers is at least three, and probably five or six orders of magnitude less than that of majority carriers.

It is impossible for the concentrations of the holes and electrons to change abruptly, since such a state of affairs would produce an infinite concentration gradient and therefore an infinite rate of diffusion of both carriers across the junction. In fact, diffusion occurs, and as it does so, an electric field is created which tends to oppose further diffusion. An equilibrium state is reached in which the field exactly cancels out the tendency to diffuse. The resulting distribution is shown in Fig. 2.1(a) in which concentration is plotted on a logarithmic scale. Note that the concentration of fixed donor and acceptor atoms does change abruptly at the metallurgical junction, since these are fixed in the crystal lattice and cannot diffuse. In the figure, the quantity p_{po} is the equilibrium concentration of holes in the p type material remote from the junction, the other symbols having similar and obvious meanings.

In a wide region near the metallurgical junction, the quantity (p–n) is several orders of magnitude less than either n_a or n_d, so that in this region the total charge density is $+qn_d$ on the n side of the junction, and $-qn_a$ on the p side. The charges in question are the

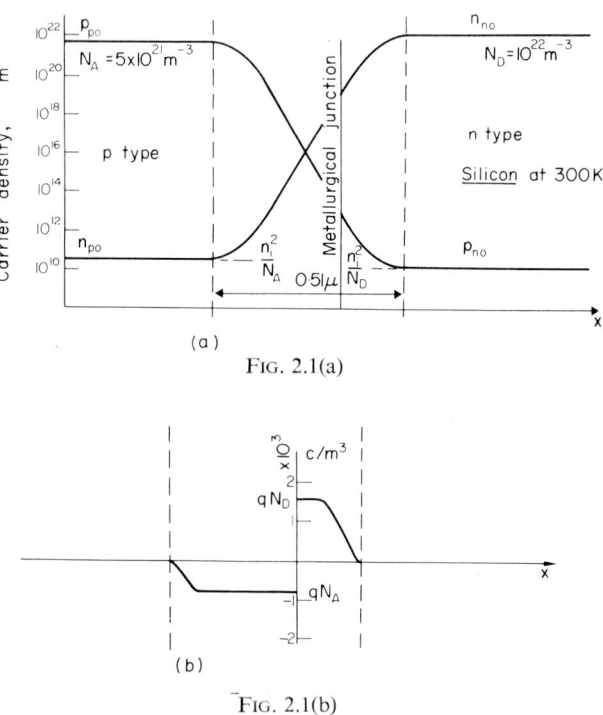

(a)

Fig. 2.1(a)

(b)

Fig. 2.1(b)

fixed residual charges upon the impurity atoms, and since this region is the only one in which the total charge density differs significantly from zero, it is known as the Space Charge region or layer. Moreover, since the region is almost completely depleted of mobile charge carriers, it is known as the Depletion Region or layer. Figure 2.1(b) illustrates the situation, and it can be seen that the total charge density falls to zero fairly rapidly at the edges of the depletion layer.

Having obtained the variation of total charge density with distance, it is a simple matter to determine the electric field and potential at any point in the semiconductor. The results are illustrated in Figs. 2.1(c) and 2.1(d). Numerical values where given are for silicon at 300 K, having $n_a = 5 \times 10^{21} m^{-3}$ and $n_d = 10^{22} m^{-3}$.

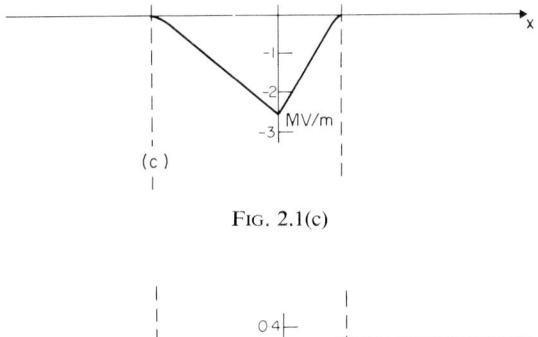

FIG. 2.1(c)

FIG. 2.1(d)

It should be noted that the potential barrier height, or contact potential, cannot be measured by means of any instrument which draws current. The reason is that any junction between dissimilar materials produces a contact potential, and the algebraic sum of these around any closed circuit must be zero. If this were not so, then current would flow, and the circuit would be self-heating, thus breaking the second law of thermodynamics. The contact potential for either silicon or germanium has a temperature coefficient of about 6 mV/K at room temperature.

The Effect of Bias

In order to explain the effect of bias applied to the pn junction diode, we make two simplifying assumptions. Firstly we assume that the contacts between the semiconductor and its terminals of metal are ohmic and of low resistance. It is possible to make such contacts, but discussion of the methods is not germane to the present topic. Secondly we assume that the voltage drops in the neutral regions of the semiconductor are negligible in comparison with the contact potential.

Under these conditions, any applied voltage will appear as a change in the height of the potential barrier. A forward bias, making the p region more positive relative to the n will evidently reduce the height of the barrier, decrease the electric field in the depletion layer, and decrease the space charge in both halves of the depletion layer: the latter it does mainly by reducing the width of the depletion layer.

Figure 2.2(a) shows the effect of a forward bias upon carrier concentrations, the x scale having been chosen to illustrate the effect in the depletion layer. Figure 2.2(b) shows the same on a scale chosen to show the effect in the neutral regions.

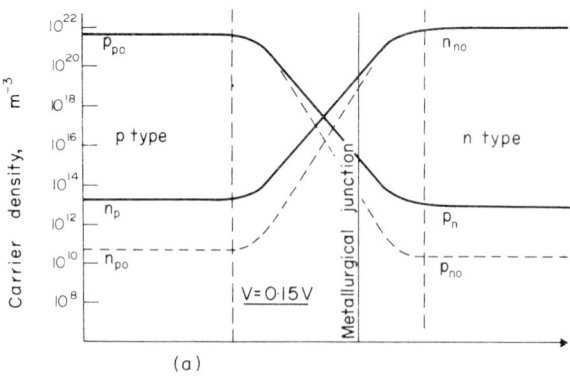

(a)

Fig. 2.2(a)

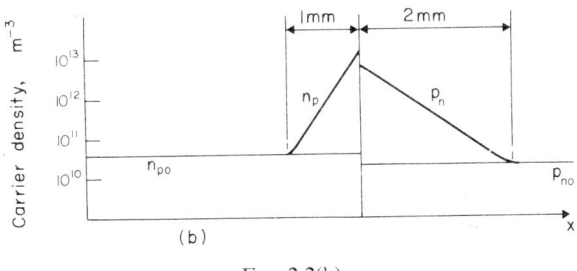

FIG. 2.2(b)

Since we have postulated a forward bias, the barrier height is reduced, and more carriers of both types can diffuse across the junction, so that the minority carrier concentrations at the edges of the depletion layer are increased above their equilibrium values. Having crossed the depletion region, the minority carriers continue to diffuse away from the junction, recombination occurring as they travel, so that the minority carrier concentration falls towards its equilibrium value as distance from the junction increases.

Changes in the majority carrier concentrations also occur, but these cannot be shown on the diagram because of the logarithmic scale. The majority carrier concentration at each edge of the depletion layer must increase by exactly the same amount as the minority carrier concentration at the opposite edge. This statement assumes that the rate of recombination within the depletion layer is insignificant, due to the small concentrations of mobile carriers there. Majority carriers required to replace those crossing the junction and also to replace those which recombine with the injected minority carriers, are supplied from the external circuit. The total current at any point is evidently the sum of the hole and electron currents, and must be independent of position. Figure 2.3 shows the current composition as a function of position. It should be noted that at any considerable distance from the junction, the current is entirely due to majority carriers.

If the applied bias is in the reverse direction, the height of the potential barrier will increase, the electric field will increase, and

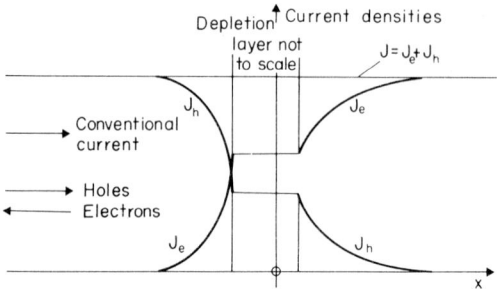

FIG. 2.3

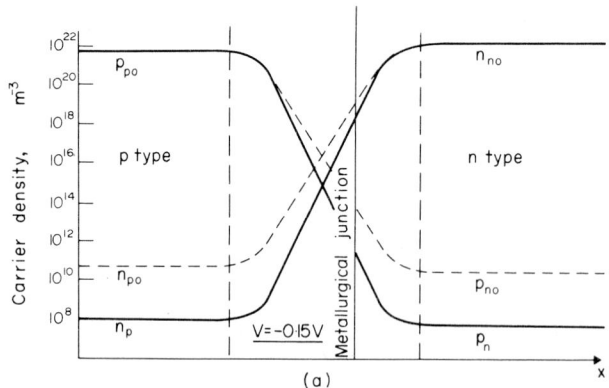

(a)

FIG. 2.4(a)

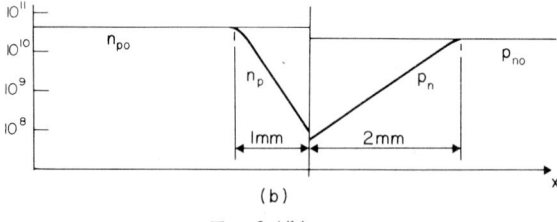

(b)

FIG. 2.4(b)

the space charge in both halves of the depletion layer will increase. Figure 2.4(a) shows the effect of reverse bias upon carrier concentrations in the depletion layer, whilst Fig. 2.4(b) shows the effect in the neutral regions.

Since the height of the potential barrier has increased, the number of majority carriers diffusing across the junction and becoming minority carriers on the opposite side has been reduced. At the same time, the number of minority carriers which are accelerated across the junction by the electric field, to become majority carriers on the opposite side has increased. Thus the minority carrier concentrations at the edges of the depletion layer are reduced by a reverse bias, and rapidly become negligible in comparison with their equilibrium values, so that further increase of reverse bias has little effect on these concentrations, and therefore upon the reverse current, which tends to become independent of voltage. Figure 2.5 shows the current distribution between holes and electrons as a function of position, for reverse bias.

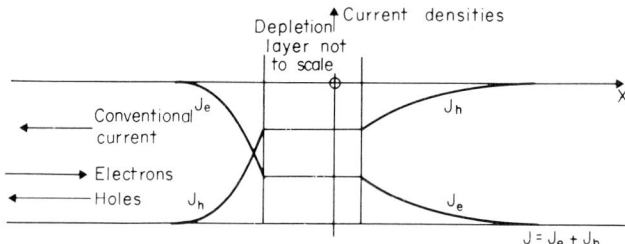

Fig. 2.5

The foregoing qualitative discussion serves to show why a pn junction has a non-linear voltage-current characteristic, and therefore rectifying properties. We shall not give a mathematical treatment here, but simply quote the results of such a treatment, which can be found in either of the two references given, and which enable practical calculations of diode behaviour in the d.c. case to be made.

The current in an idealized diode is given by

$$I = I_o (e^{qV/kT} - 1) \tag{8}$$

where

I_o is the saturation reverse current

V ,, ,, applied terminal voltage

q ,, ,, electronic charge

T ,, ,, temperature (K)

k ,, ,, Boltzmann's constant.

For a given diode and value of V, the current will depend strongly upon the temperature, mainly because I_o varies rapidly with temperature.

A practical diode will depart somewhat from the idealized behaviour of equation 8, in the main for two reasons. Firstly, the saturation reverse current, particularly in silicon, is considerably greater than that predicted by the foregoing considerations. This is because surface leakage, and the effects of small amounts of impurity accidentally present, mask the idealized effects. Secondly, equation 8 neglects the effect of the ohmic resistance of the neutral regions, which is by no means negligible at any but the smallest values of current. A more accurate, but for obvious reasons less useful, equation is

$$I = I_o (e^{q(V-Ir)/mkT} - 1) \tag{9}$$

Fortunately, for small signal low power circuits, the simpler equation does all that is required in most cases. In many further cases, it is often sufficient to take into account the diode resistance assuming it to be constant.

Figure 2.6 shows typical diode characteristic curves for germanium and silicon and it can be seen that a small change in voltage across the junction is sufficient to produce a large change in current.

The rectifying properties of the junction are very clear from Fig. 2.6 and such diodes are very commonly used for demodulation purposes in radio apparatus, but it is also clear that a simple diode junction is not in itself capable of amplification. However, if two junctions are arranged as shown in either of the alternatives given in Fig. 2.7 current, voltage and power amplification become

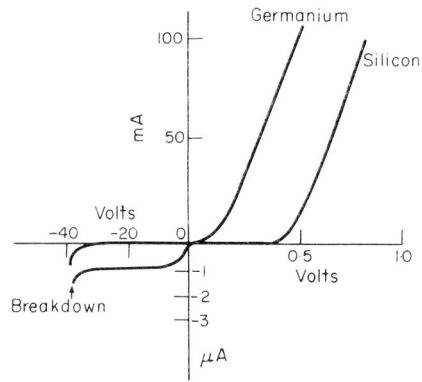

FIG. 2.6

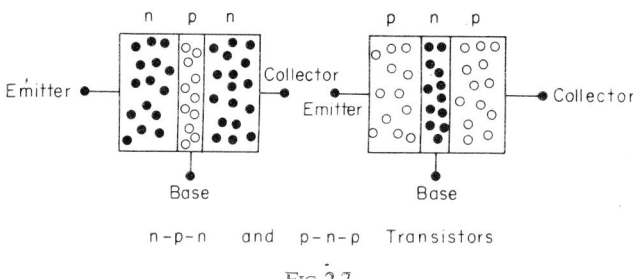

n–p–n   and   p–n–p   Transistors

FIG 2.7

possible and the device is known as a transistor. If "p" material is sandwiched between two pieces of "n" material the device is known as an n–p–n transistor, and alternatively as a p–n–p transistor if "n" material is between two pieces of "p" material. In each case the middle section is made very thin so that a forward bias across the junction between emitter and base, which will cause a copious flow of carriers, will ensure that most of the carriers will cross the base region and travel into the collector region. As has been seen above with regard to a diode junction a small change in potential in the forward direction will produce a large change in current and, because the base-collector junction is reverse biased, its resistance is high and thus the large change in current also produces a large change in voltage. Hence amplification is obtained. Figure 2.8 shows this action diagrammatically with values typical for a silicon transistor and it can be seen that a change of 0·02 volts in the base to emitter voltage gives a change of 0·995 mA in the collector current for a change of 0·005 mA in the base current. The ratio of these currents is the *short-circuit forward current ratio* and here has a value of about 200. It is often given the symbol h_{fe}.

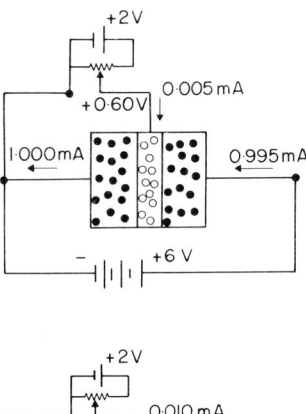

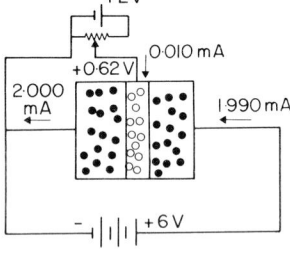

FIG. 2.8

If a load resistor is connected in series with the collector then the consequent current changes detailed above will give rise to voltage and power changes in the resistor. Hence voltage and power amplification will be obtained.

The basic principles described above of employing more than one junction may be developed in a very large number of ways resulting in a wide variety of semiconductor devices. Several of these are illustrated on the following pages where the construction, electrical characteristics and major applications for the more common devices are given. The opportunity of varying the material, doping ratio, thickness, area, temperature, etc., makes it possible to design devices having almost any required characteristic and hence devices are on the market to suit almost every need of the circuit designer. Reference to makers' lists will show that every device given on the following pages is available in a large variety of forms having a wide range of ratings for voltage, current and frequency range.

References:

An Introduction to Physical Electronics by A. H. W. BECK and H. AHMED.
Physics of Solid State Electronics by J. N. SHIVE.

Comparative Characteristics of Active Devices

Characteristic	Vacuum tube	Small-signal transistor	High-power transistor	Junction fet	Mosfet
Input impedance	High	*	Very low	High	Very high
Output impedance	High	*	Low/moderate	Hign	High
Noise	Low	Low	Moderate	Low	Unpredictable
Warm-up time	Long	Short	Short	Short	Short
Power consumption	Large	Small	Moderate	Very small	Very small
Aging	Appreciable	Low	Low	Low	Moderate
Reliability	Poor	Excellent	Very good	Excellent	Very good
Overload sensitivity	Excellent	Good	Fair	Good	Poor
Size	Large	Small	Moderate	Small	Small

* Impedance depend on circuit arrangement:

	Input impedance	*Output impedance*
For common base	Low (10's of ohms)	High (megohms)
For common emitter	Medium (kilohms)	Medium (10's of kilohms)
For common collector	High (100's of kilohms)	Low (100's of ohms)

Major Semiconductor Components

NAME OF DEVICE	CIRCUIT SYMBOL	COMMONLY USED JUNCTION SCHEMATIC	ELECTRICAL CHARACTERISTICS	MAJOR APPLICATIONS	ROUGHLY ANALOGOUS TO:
Diode or Rectifier	ANODE / CATHODE	ANODE / CATHODE	Conducts easily in one direction, blocks in the other	Rectification, Blocking, Detecting, Steering	Check valve, Diode tube, Gas diode
Avalanche (Zener) Diode	ANODE / CATHODE	ANODE / CATHODE	Constant voltage characteristic in negative quadrant	Regulation, Reference, Clipping	V–R tube
Integrated Voltage Regulator (IVR)	IVR		Programmed to desired V_{21} by two resistors	Shunt voltage regulator, Reference element, Error modifier, Level sensing, Level shifting	Avalanche Diode
Tunnel Diode	POSITIVE ELECTRODE / NEGATIVE ELECTRODE	POSITIVE ELECTRODE / NEGATIVE ELECTRODE	Displays negative resistance when current exceeds peak point current I_p	UHF converter, Logic circuits, Microwave circuits, Level sensing	None
Back Diode	ANODE / CATHODE	ANODE / CATHODE	Similar characteristics to conventional diode except very low forward voltage drop	Microwave mixers and low power oscillators	None

Device	Characteristic	Applications	Vacuum Tube Equivalent
Diode	Rapidly increasing current above rated voltage in either direction (I vs VOLTAGE)	Transient voltage suppression and arc suppression	Thyrite Two avalanche diodes in inverse-series connection
n-p-n Transistor (COLLECTOR, BASE, EMITTER, I_C, I_B)	Constant collector current for given base drive (I_C vs $V_{COLLECTOR}$ (+); IB5, IB4, IB3, IB2, IB1)	Amplification Switching Oscillation	Pentode Tube
p-n-p Transistor (COLLECTOR, BASE, EMITTER, I_C, I_B)	Complement to n-p-n transistor ($V_{COLLECTOR}$ (-), $I_{COLLECTOR}$ (-); IB1, IB3, IB4, IB5)	Amplification Switching Oscillation	None
Photo Transistor (COLLECTOR, BASE, EMITTER, I_B)	Incident light acts as base current of the photo transistor (I COLLECTOR vs V_{CE}; H4, H3, H2, H1)	Tape readers Card readers Position sensor Tachometers	None
Unijunction Transistor (UJT) (BASE 2, BASE 1, EMITTER, I_V)	Unijunction emitter blocks until its voltage reaches V_p then conducts (VOLTAGE BETWEEN EMITTER & BASE 1 vs EMITTER I_e; V_p)	Interval timing Oscillation Level Detector SCR Trigger	None

Major Semiconductor Components

NAME OF DEVICE	CIRCUIT SYMBOL	COMMONLY USED JUNCTION SCHEMATIC	ELECTRICAL CHARACTERISTICS	MAJOR APPLICATIONS	ROUGHLY ANALOGOUS TO
Complementary Unijunction Transistor (CUJT)			Functional complement to UJT	High stability timers Oscillators and level detectors	None
Programmable Unijunction Transistor (PUT)			Programmed by two resistors for V_p, I_p, I_v. Function equivalent to normal UJT	Low cost timers and oscillators Long period timers SCR trigger Level detector	UJT
Silicon Controlled Rectifier (SCR)			With anode voltage (+), SCR can be triggered by I_g, remaining in conduction until anode I is reduced to zero	Power switching Phase control Inverters Choppers	Gas thyratron or ignitron
Complementary Silicon Controlled Rectifier (CSCR)			Polarity complement to SCR	Ring counters Low speed logic Lamp driver	None
Light Activated SCR* (LASCR)			Operates similar to SCR, except can also be triggered into conduction by light falling on junctions	Relay Replacement Position controls Photoelectric applications Slave flashes	None

Device	Symbol	Construction	Characteristic	Operation	Applications	Equivalent
Silicon Controlled Switch* (SCS)	ANODE / CATHODE GATE / ANODE GATE / CATHODE	ANODE / CATHODE GATE / ANODE GATE / CATHODE	I vs V_{ANODE}	Operates similar to SCR except can also be triggered on by a negative signal on anode-gate. Also several other specialized modes of operation	Logic applications, Counters, Nixie drivers, Lamp drivers	Complementary transistor pair
Silicon Unilateral Switch (SUS)	ANODE / GATE / CATHODE	ANODE / GATE / CATHODE	$I_{ANODE} (+)$ vs $V_{ANODE} (+)$	Similar to SCS but zener added to anode gate to trigger device into conduction at ~ 8 volts. Can also be triggered by negative pulse at gate lead.	Switching Circuits, Counters, SCR Trigger, Oscillator	Shockley or 4-layer diode
Silicon Bilateral Switch (SBS)	ANODE 2 / GATE / ANODE 1	R_B GATE ANODE 2 / ANODE 1 R_B	$I_{ANODE\ 2}$ vs $V_{ANODE\ 2(-)}$ / $V_{ANODE\ 2(+)}$	Symmetrical bilateral version of the SUS. Breaks down in both directions as SUS does in forward.	Switching Circuits, Counters, TRIAC Phase Control	Two inverse Shockley diodes
Triac	ANODE 2 / GATE / ANODE 1	ANODE 2 / GATE / ANODE 1	I vs $V_{ANODE\ 2(+)}$ / $V_{ANODE\ 2(-)}$	Operates similar to SCR except can be triggered into conduction in either direction by (+) or (-) gate signal	AC switching, Phase control, Relay replacement	Two SCR's in inverse parallel
Diac Trigger	ANODE 2 / ANODE 1	c p c	I vs V	When voltage reaches trigger level (about 35 volts), abruptly switches down about 10 volts.	Triac and SCR trigger, Oscillator	Neon lamp

CHAPTER 3

Resistance, Capacitance and Inductance

IN CHAPTER 1 we have considered certain electrical properties of materials without in any way considering how the material might be used. We shall now consider the use of material in components designed to have a particular property.

The greater part of an electrical assembly is made up of simple parts each of which must have an electrical resistance, capacitance and inductance. Often, in a particular part of the circuit, one only of these is necessitated by the design of the circuit, the others being regarded as imperfections. On the other hand there are some circuits where the behaviour of a component depends almost entirely on its possessing simultaneously appropriate values of two or all three of these parameters. However if all the component parts of a circuit consisted of pure resistors, capacitors and inductors, it would obviously be possible to make the circuit equivalent to one consisting of real components which had inherent combinations of all three parameters and we will thus first consider these pure circuit elements.

Pure resistor symbol O———-√√√√√√———-○

FIG. 3.1

Figure 3.1 gives the circuit symbol for a pure resistor. A value put against it is in ohms, symbol Ω, and signifies that the resistance between its ends has that value. An ideal resistor does not change its value with temperature or for any other reason

34

and will carry an infinite current without being destroyed. It has, of course, by definition no capacitance or inductance.

Pure capacitor symbol   o———————||————————o

FIG. 3.2

Figure 3.2 gives the circuit symbol for a pure capacitor. Its value is in farads (or more commonly, for convenience, microfarads), symbol F (or μF). It also will not change in value however its environment changes, it will stand the application of an infinite potential difference across its terminals and hence will store an infinite amount of electricity. Its dielectric has infinite resistivity so that the steady application of a p.d. produces no current and when current flows into it or out of it there is no loss of energy. There can be no leakage of energy either, so that when charged to a given p.d. the potential between the terminals will remain the same indefinitely after the supply from which it was charged has been disconnected from it.

Pure inductor symbol   o———ↄↄↄↄↄↄↄↄ———o

FIG. 3.3

Figure 3.3 gives the circuit symbol for a pure inductor. Its value is in henries where the inductance L is given by

$$L = \frac{\text{flux} \times \text{turns}}{\text{amps}}$$

In the perfect component *all* the flux produced by the turns always passes through *all* the turns. The conducting material of the turns has zero resistivity and the surroundings of the coil do not cause a loss of energy when a changing magnetic field is established in them. The coil can carry an infinite current without harm, thus a perfect inductor can also store an infinite amount of electrical energy.

We must now consider how nearly real components approach the ideal.

Resistors may be made from a very large number of materials. If a resistor is made of a rod of material of cross-section A, length l and resistivity ρ then its resistance R is given by

$$R = \frac{\rho l}{A} \text{ ohms.}$$

Thus it can be seen that for any value whatever of ρ the ratio l/A can be made appropriate to any required value of R. It is obvious that if ρ is very great and R is required to be small then the length will be very short and the cross-section large. Conversely for a large R from material with a small ρ the cross-section may be inconveniently small and the length still very great. When considerations of bulk, cost, ease of manufacture, stability and small inductance and capacitance are taken into account, the choice of material is considerably narrowed, and the majority of resistors are made from either metal, wire or moulded carbon. Wire-wound resistors are more stable and may be more accurately made to a given value than carbon ones. The wire is usually one of the special alloys developed for the purpose such as Nichrome, Eureka or Manganin which have low temperature coefficients of resistance. Since a resistor always dissipates energy it is bound to get hot and the cross-section and length of the wire must be chosen not only to give the correct value of R but also to give sufficient surface area for the dissipation of the heat produced at the maximum current the resistor is allowed to carry. If the rate of energy being dissipated is W watts then

$$W = I^2 R \text{ watts}$$

and if the rate of dissipation per unit surface area of the wire is known it is simple to calculate the required surface area.

If the resistance wire comprising the resistor is simply wound as a coil on a former it will have some inductance which may be considerable. However if two equal coils of the wire were wound together on the same former and connected so that they were in series but also so that the current went one way round one coil and the opposite way round the other then negligible net flux would be produced and the resultant resistor would be said to be *non-inductively* wound.

The carbon resistor is made quite differently, being in effect just a short rod of the material which can be pure carbon compressed into a rod or else carbon mixed with a clay similarly moulded. Here there is very little inductance since the resistor effectively consists of only part of one turn. It will be clear that this construction produces very little surface area for a given volume of material compared with that possible with a long wire, hence the power that can be dissipated from any practical carbon resistor is low, not more than 5 W and usually a good deal less. The temperature coefficient of carbon, about -0.0009, is considerably less than that for many metals but is nothing like as low as that for the special resistance-wire materials. This, coupled with cooling difficulties which make it necessary to run the resistors at a higher temperature than wire-wound resistors run at, makes the ohmic value of a carbon resistor somewhat variable in use. In addition, the moulding method does not produce a completely stable product so that the value of resistance can change with use. There are, nevertheless, a great number of circuit situations where the carbon resistor is absolutely satisfactory because variations of ± 10 per cent do not matter, the power to be dissipated is low, and the small size is a great advantage.

A third type of resistor, mainly of high or very high ohmic value consists of a thin film of conducting material deposited on a support made of insulating material. These components are called "metal film" or "oxide film" resistors except when the conducting film is of carbon when the term "cracked carbon" resistor is used. They can be manufactured to a higher degree of accuracy than moulded carbon resistors but not nearly so precisely as the wire-wound type as can be seen from Table 4 which gives a comparison of properties and accuracy of manufacture of several common types.

Figure 3.4 shows a common construction for some of the main types given in Table 4 but there is considerable variation between the products from different manufacturers of these.

Variable resistors are constructed from the same materials as the fixed ones, but in general they are of lower accuracy and stability and are also subject to wear if, as is usual, they incorporate a sliding contact.

TABLE 4

Resistor	Manu-facturing accuracy %	Stability %	Temp. coeff. °C	Max. temp. °C	Power watts
Precision wire-wound	±0·01	±0·01	+0·00002	65	$\frac{1}{10}$
Wire-wound	±1	±1	±0·0002	300	3–300
Metal film	±1	±1	+0·00025	150	$\frac{1}{2}$–2
Cracked carbon	±1	±2	−0·0005	150	$\frac{1}{8}$–1$\frac{1}{2}$
Moulded carbon	±20	±25	±0·001	120	$\frac{1}{4}$–2$\frac{1}{2}$

Practical capacitors consist of conducting electrodes connected to the two terminals of the component and separated by a dielectric. It is not necessary that the electrodes shall be metallic since a conducting fluid can act just as well, but the metallic electrode is the commonest. The dielectric can be any convenient insulating material, including gases, or alternatively appropriately spaced electrodes may be enclosed in an evacuated chamber. The choice of dielectric is influenced mainly by its relative permittivity, its maximum allowable potential gradient, its leakage current when incorporated in the component, its power loss when subjected to an alternating electric field and its cost.

Table 5 gives some of the properties of common dielectrics used in capacitor construction.

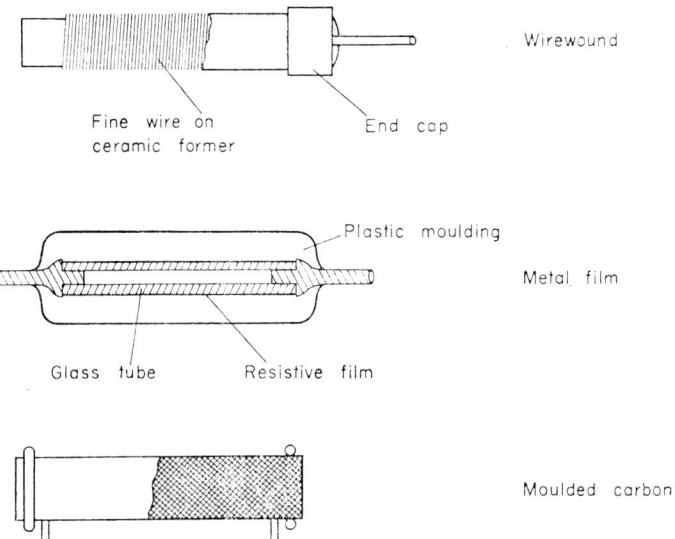

Fine wire on ceramic former
End cap
Wirewound

Plastic moulding
Metal film

Glass tube
Resistive film

Moulded carbon

FIG. 3.4

TABLE 5

Dielectric	Relative permittivity	Time to lose half of charge	Stability %	Manufacturing accuracy %	Power factor
Vacuum	1	Days	—	—	—
Air	1·0006	Days	±0·01	±0·01	0·00001
Polystyrene	2·6	Days	±0·5	±20	0·0005
Waxed paper	5	Hr	±5	±20	0·01
Mica	6·5	Hr	±1	±10	0·002
Ceramic	100	Hr	±1	±10	0·001
High permittivity ceramic	1000	Min	±20	±20	0·01
Electrolytic	—	Sec	±10	±20	0·05

Figure 3.5 shows the construction of two typical capacitors used for many purposes.

Variable capacitors are frequently constructed as shown in Fig. 3.6. In general for this the dielectric is air, although it is quite possible to insert sheets of any other dielectric between the stationary and moving plates. Another common type of small

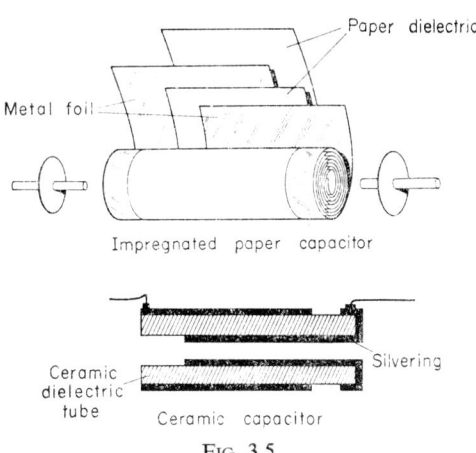

Impregnated paper capacitor

Ceramic capacitor

FIG. 3.5

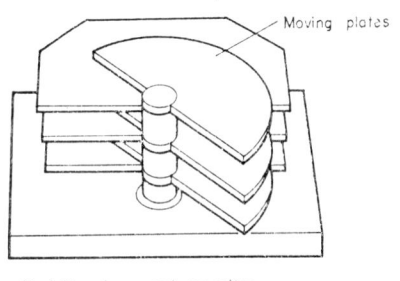

Variable air spaced capacitor

FIG. 3.6

variable capacitor is constructed as shown in Fig. 3.7 where the dielectric is mica and the capacitance is varied by altering the spacing between the plates.

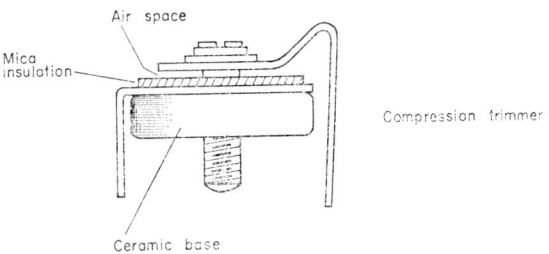

FIG. 3.7

Practical inductors are constructed in many different ways and it is difficult to generalize about them. They are essentially just a coil of wire wound on a former or support which may or may not be made from a magnetic material. For a given value of inductance it will clearly be possible to produce a smaller component by winding the wire on a core having a high magnetic permeability than that produced by winding it on a non-magnetic core, but several other considerations such as hysteresis and eddy loss and also variation of permeability with current and temperature may make a non-magnetic core necessary. On the other hand certain types of variable inductor, and transducers working on an inductance principle, make use of these variations of properties. Mention here should be made of "Ferrites" which are materials having a high permeability and low loss at medium and high frequencies such as are used in the operation of radio and television systems. The various grades of the nickel–zinc ferrites, for example, are suitable for inductor cores for use at frequencies from 1 kc/s up to 50 Mc/s.

Figure 3.8 shows an example of an iron cored inductor. It is not possible to give a useful table of properties since these depend so much on the frequency at which the inductor is being used.

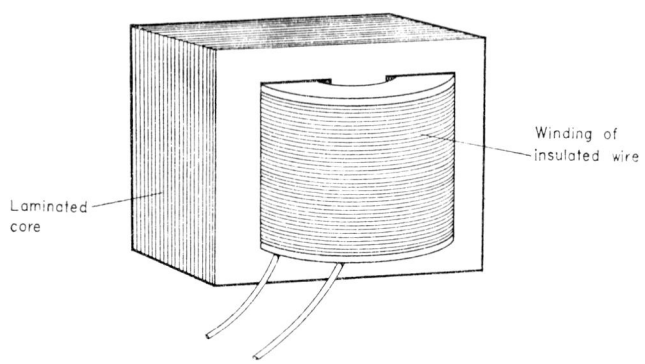

Winding of
insulated wire

Laminated
core

FIG. 3.8

It is possible to design a low frequency inductor having an iron core and working at 100 c/s so that its power factor is about 0·01, but at higher frequencies the same inductor would have a much greater power loss. For the high frequencies air or ferrite cores would be used and the power factor could then be less than 0·01.

The components described so far are known as *discrete* components and, whilst they are reasonably compact, they would be far too large for use in apparatus like digital computers, portable tape recorders, transistor radios, etc. Also for very high frequency apparatus the lengths of the connections necessary to join up discrete components would put an unsatisfactory upper limit on the frequencies acceptable. Diodes, transistors and other semiconductor devices together with their circuit associated resistors and capacitors are therefore frequently manufactured in very much reduced size by means of *integrated circuit* techniques. For low power use the active area of a semiconductor device is very small

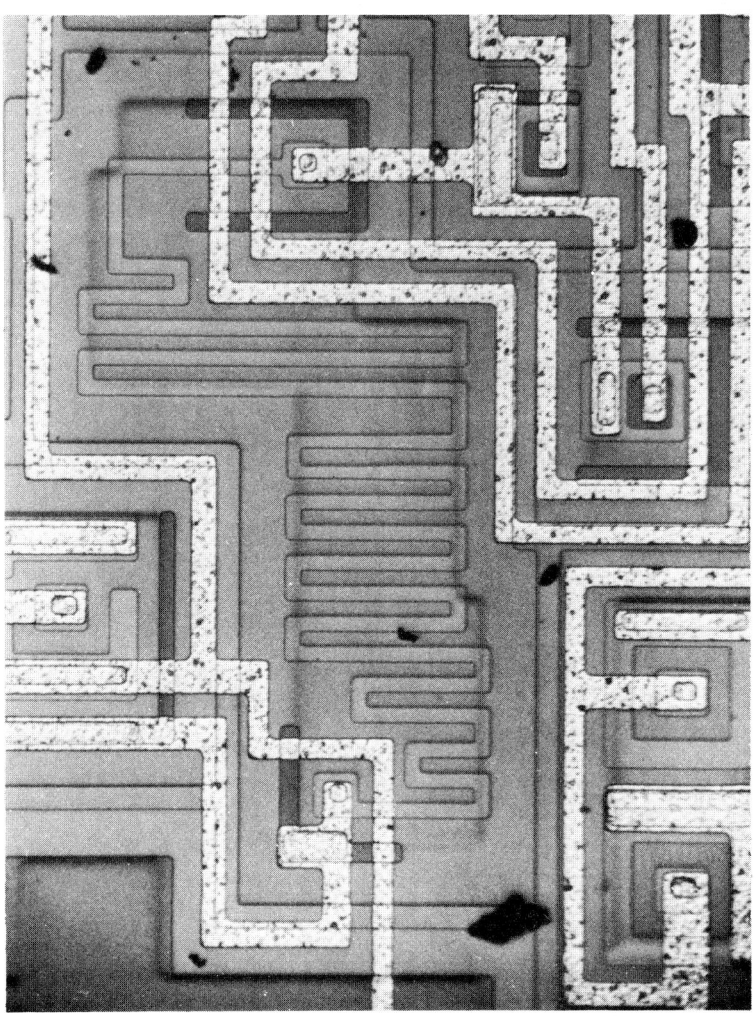

$\times$ 300

FIG. 3.9

and the dimensions of the device proper may be only a few micro-metres each way, hence if most of the bulk of the connections can be eliminated and if one device can be made contiguous with the next, or with a component such as a resistor or a capacitor, an extremely high packing density can be achieved giving as many as 100 components per square millimetre as shown in the photograph Fig. 3.9.

The components and devices are built up on a small slice of silicon about 0·2 mm thick measuring about 2 mm by 3 mm, the building-up process consisting of a series of photographically-produced masking stages interspersed with etching and diffusion whereby the appropriate parts of each component are constructed in layer form as shown in Fig. 3.10.

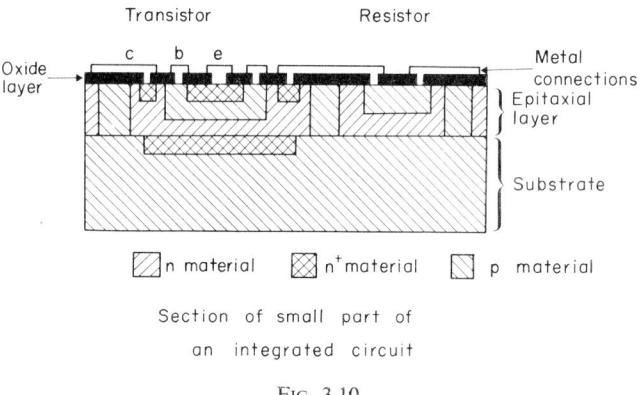

Section of small part of
an integrated circuit

FIG. 3.10

Because slices of silicon about 5 cm in diameter can be produced quite readily, it is possible to cover the surface of a slice of this size with a very large number of repetitions of a complete circuit so that the cost of each circuit can be kept to a very low figure, the circuits being separated from one another after manufacture by cutting up the large slice with a diamond.

Further reductions in size can be achieved by using metal oxide semiconductor transistors to perform as resistors, since by this means the area occupied is only one hundredth of that required for a resistor produced in an integrated circuit by the conventional diffusion method. Considerable ingenuity has been used by manufacturers in this way to achieve the very smallest integrated circuits, with consequent decrease in price and increase in portability and reliability.

CHAPTER 4

Resistive Circuits

THE SIMPLEST possible resistive circuit consists of a single ideal resistor of value R ohms connected to a steady source of e.m.f. E volts as shown in Fig. 4.1. Application of Ohm's law to this circuit gives a value of steady current

$$I = \frac{E}{R}$$

which follows from Ohm's original statement that I would be proportional to E if R were kept constant which is true if the resistor is ideal.

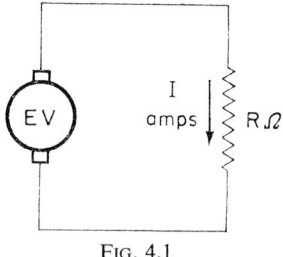

FIG. 4.1

Now it is also possible to connect a constant current source to the resistor as shown in Fig. 4.2 and it is clear by application of the above equation that the potential difference across the resistor will be given by $V = IR$ where V takes the place of E. Whether the resistor is connected to a constant voltage or a constant current

46

source it is always possible to arrange that conditions in the resistor are as required and the quantity of electricity passing round the circuit in time t seconds will be

$$Q = It \text{ coulombs}$$

which is closely analogous to water flowing in a pipe where, clearly, the amount of water passing any point is the product of the rate of flow and the time, provided that the rate of flow is constant.

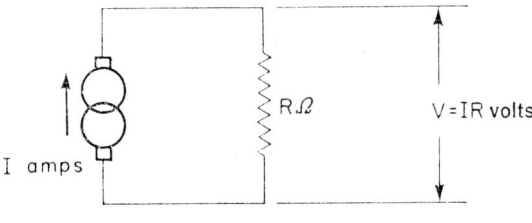

FIG. 4.2

Whilst Q is of interest in certain cases the rate of energy dissipation or power, P, is more important in most practical circumstances because it results in temperature rise in the resistor which, as we have seen in Ch. 3, changes its ohmic value.

Power in the resistor connected as in Figs. 4.1 and 4.2 is given by

$$P = VI$$

which again has a water flow analogy where pressure multiplied by flow gives the rate of work being performed (i.e. lb/ft^2 × ft^3/sec = lb ft/sec = work/sec).

It is also convenient, using $V = IR$, to write

$$P = \frac{V^2}{R} = I^2 R$$

Except in the very simplest circuits, resistors are used in combination with one another and these are basically either in series as in Fig. 4.3 or in parallel as in Fig. 4.4. These combinations may themselves be connected in parallel or series thus building up complicated networks. The analysis and solution of the latter is usually much simplified by breaking them up into simpler component parts.

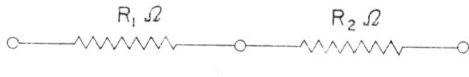

Fig. 4.3

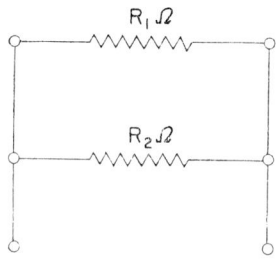

Fig. 4.4

Let us then consider two resistors of value R_1 and R_2 connected in series as in Fig. 4.5. The combination of the two is equivalent to one resistor of value $R = R_1 + R_2$ connected to the source of electricity and hence the current flowing will be given by

$$I = \frac{V}{R} = \frac{V}{R_1 + R_2}$$

and this current will be the same for R_1 and R_2. Therefore the potential differences across R_1 and R_2 will be respectively

$$V_1 = IR_1 \quad \text{and} \quad V_2 = IR_2$$

The arrangement of resistors shown in Fig. 4.6 is often called a potential-divider where the input potential V_{in} is the same as V

and the output potential V_{out} is V_2. When used in this way the network is said to have a transmission factor given by

$$\frac{V_{\text{out}}}{V_{\text{in}}} = \frac{IR_2}{I(R_1 + R_2)} = \frac{R_2}{R_1 + R_2}$$

It is also possible to make the ratio between R_1 and R_2 variable whilst keeping their sum the same, in which case we get the arrangement shown in Fig. 4.7 which is often called a potentiometer but this must not be confused with the precision measuring

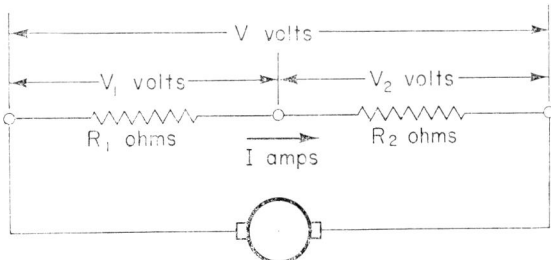

FIG. 4.5

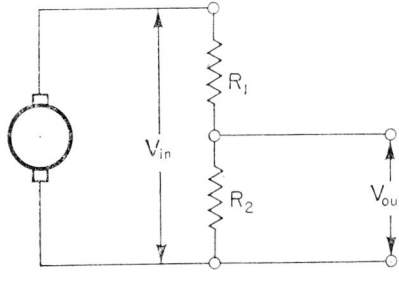

FIG. 4.6

instrument bearing the same name although the latter does depend for its operation on the same basic circuit as will be shown later in this chapter.

The alternative connexion for two resistors R_1 and R_2 is in parallel as shown in Fig. 4.8. Here it is obvious that the potential difference across the resistors is the same for each. The currents in R_1 and R_2 respectively are thus:

$$I_1 = \frac{V}{R_1} \quad \text{and} \quad I_2 = \frac{V}{R_2}$$

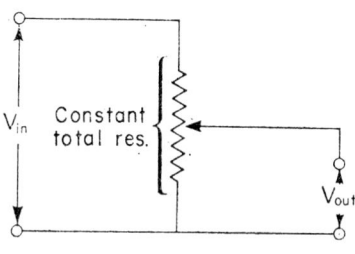

FIG. 4.7

giving a total current

$$I = I_1 + I_2 = \frac{V}{R_1} + \frac{V}{R_2} = V\left(\frac{1}{R_1} + \frac{1}{R_2}\right).$$

Now if the parallel combination of R_1 and R_2 were to be replaced by a single resistor of value R ohms then

$$I = \frac{V}{R} = V\left(\frac{1}{R}\right)$$

$$\therefore \quad \frac{1}{R} = \left(\frac{1}{R_1} + \frac{1}{R_2}\right)$$

and

$$R = \frac{R_1 R_2}{R_1 + R_2}$$

so that the effective value of resistance connected to the source is given by the product of the two actual resistors divided by their sum.

This method of determining the equivalent resistance can be extended to any number of parallel resistors thus giving

$$R_{\text{equiv.}} = \frac{1}{1/R_1 + 1/R_2 + \dots + 1/R_n}$$

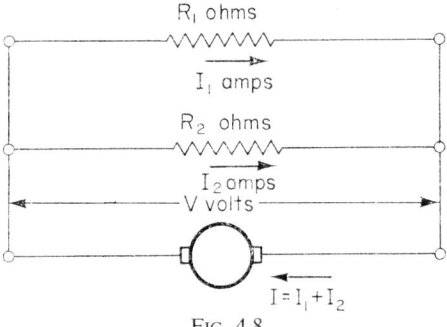

FIG. 4.8

If the applied potential difference is known there is no difficulty in determining the individual currents in the resistors, but if the circuit is connected to a constant current source and it is necessary to determine the individual currents we can proceed as follows:

$$R_{\text{equiv.}} = \frac{R_1 R_2}{R_1 + R_2}$$

$$\therefore \quad V = I\left(\frac{R_1 R_2}{R_1 + R_2}\right)$$

$$\therefore \quad I_1 = \frac{V}{R_1} = \frac{I}{R_1}\left(\frac{R_1 R_2}{R_1 + R_2}\right) = I\left(\frac{R_2}{R_1 + R_2}\right)$$

similarly $I_2 = I[R_1/(R_1 + R_2)]$.

It may be noted that the distribution factors $[R_1/(R_1 + R_2)]$ and $[R_2/(R_1 + R_2)]$ are the same as the transmission factors for the potential divider connexion but it is important to remember that whereas in the latter case the factor having numerator R_1 gives the potential across R_1, the distribution factor with numerator R_1 gives the current in R_2.

Before proceeding to the solution of more elaborate resistive networks, we must next consider real as opposed to ideal sources. In general a real source will consist of an arrangement for producing an e.m.f. and various connexions which will have resistance

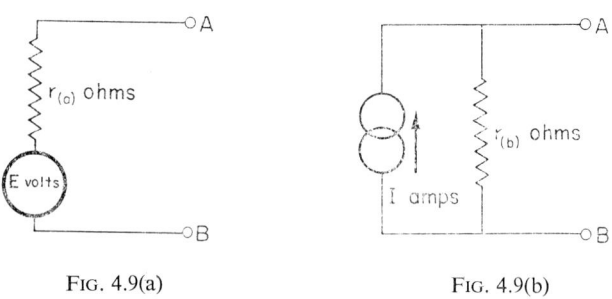

FIG. 4.9(a)   FIG. 4.9(b)

and may be in parallel or in series with the e.m.f. It is possible by experiment to determine the relation between the terminal p.d. and the current output from a real source and, if this is linear, hence to determine the value of one single fixed resistor and of one constant e.m.f. in series, or alternatively one constant current with one single fixed resistor in parallel, which will in either case correctly represent the actual source so far as terminal p.d. and current are concerned. Most real sources are in fact non-linear over some part at least of the output characteristic but in many cases the non-linearity is not great and in others the characteristic is straight over the practical working portion of the curve so that representation of the source by a fixed e.m.f. or current and a fixed resistor give close approximations to actuality.

Figure 4.9(a) and (b) show the diagrammatic representations

of such real, linear sources where the terminals to which loads would be applied are A and B in each case. If a voltmeter which drew no current were connected between A and B it would read E volts for circuit (a) and $r_{(b)}I$ volts for circuit (b) and if an ammeter which had zero internal resistance were connected across the same terminals in place of the voltmeter it would read $E/r_{(a)}$ amps for circuit (a) and I amps for circuit (b), the latter being true because *all* the constant current I would have to go through the ammeter and none through the resistor because the ammeter has zero resistance. From these results, if (a) and (b) are meant to represent the *same* actual source, we get the following relations:

From voltmeter test: $\quad E = r_{(b)}I$

From ammeter test: $\quad \dfrac{E}{r_{(a)}} = I$

Thus if the two circuits are to represent the same source $r_{(b)} = r_{(a)} = r$ and if the actual source is linear, or can be assumed to be so, then the output voltage–current relation for circuit (a) or (b) will be the same as that for the actual source for all values of current.

Since an actual source can always be represented by circuit (a) or circuit (b) it is necessary to consider which is the best way in any particular circumstance. In some instances there is not much to choose between the two but it is often the case that the effective resistance, R, of the load is either much greater or much less than r. Remembering that the load is connected across A–B in any case, it can be seen from Fig. 4.10(a) that with $R \gg r$ the p.d. across R is

$$V_L = E\left(\frac{R}{R+r}\right) \simeq E \quad \text{because} \quad R \simeq (R+r)$$

and that from Fig. 4.10(b) with $R \ll r$

$$V_L = I\left(\frac{Rr}{R+r}\right) \simeq IR \quad \text{because} \quad (R+r) \simeq r$$

Both these results are, of course, approximations and will be
in error by an amount depending on the inequality of R and r;
the greater the inequality the less the error. Whether or not this
error is important depends on the nature of the load. For instance
a two per cent error in calculating the current through an electric
fire element would not matter whereas a similar error in the current
through a voltmeter would, for a high grade instrument, be outside
the limits of accuracy normally specified for such instruments.

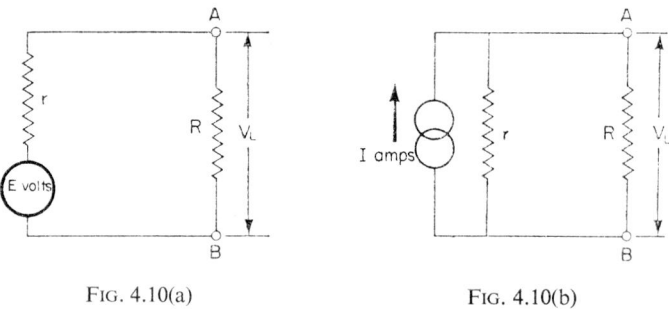

FIG. 4.10(a)   FIG. 4.10(b)

Another quite different consideration, not involving approximation, influencing the choice of source representation is the form
of connexion of any complex load consisting of several resistors
which is connected to terminals A and B. If the complex load
can be represented by an all series circuit then source of type (a)
will involve less algebra as will source (b) when the load is all
parallel.

We are now ready to consider the solution of some simple networks when connected to a source. The relations worked out at
the beginning of this chapter for solving series or parallel arrangements of resistors can be used for a class of quite complex networks in which for every point where the current divides into two
portions there is another point where these portions combine
again. A network of this type is shown in Fig. 4.11(a) connected
to a voltage source of e.m.f. $E = 25$ V and internal resistance

$r = 1\Omega$. The solution of this network using the series parallel method is as follows:

(1) Start at the part of the network furthest from the terminals A and B, i.e. the part having $3\ \Omega$ in parallel with $6\ \Omega$.

(2) Successively determine the equivalent resistance values of resistors in parallel or in series until the network has been reduced to one equivalent resistor across A–B.

(3) Determine the current through this resistor when supplied by the given source.

(4) Working forward through the network determine the potential difference across, and the current in, each branch.

The numerical working is as follows and it is convenient to redraw the circuit as shown after each reduction of a network branch.

Fig. 4.11(b)   Equivalent resistance from G–H

$$= \frac{3 \times 6}{3 + 6} = \frac{18}{9} = 2\ \Omega.$$

Fig. 4.11(c) ∴ Equivalent resistance from E–H

$$= 10 + 2 = 12\ \Omega.$$

Fig. 4.11(d)   Equivalent resistance from E–F

$$= \frac{4 \times 12}{4 + 12} = \frac{48}{16} = 3\ \Omega.$$

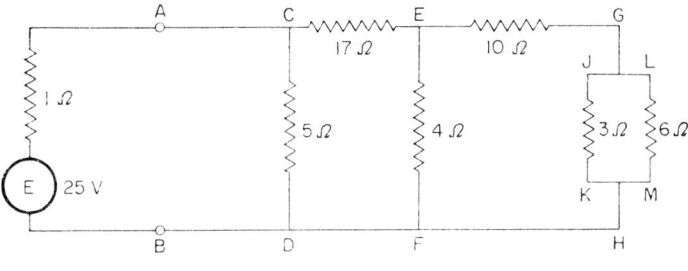

Fig. 4.11(a)

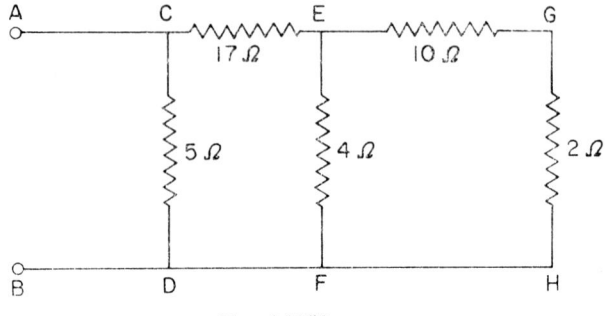

Fig. 4.11(b)

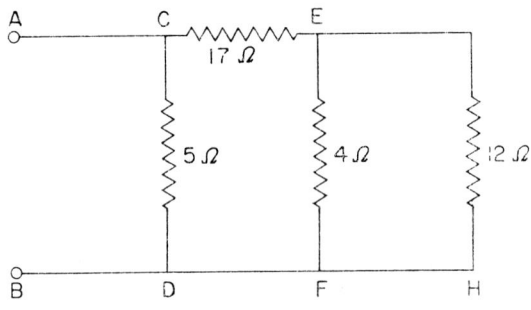

Fig. 4.11(c)

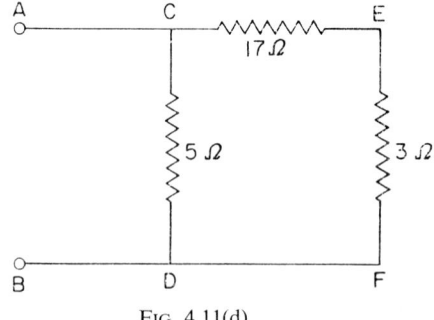

Fig. 4.11(d)

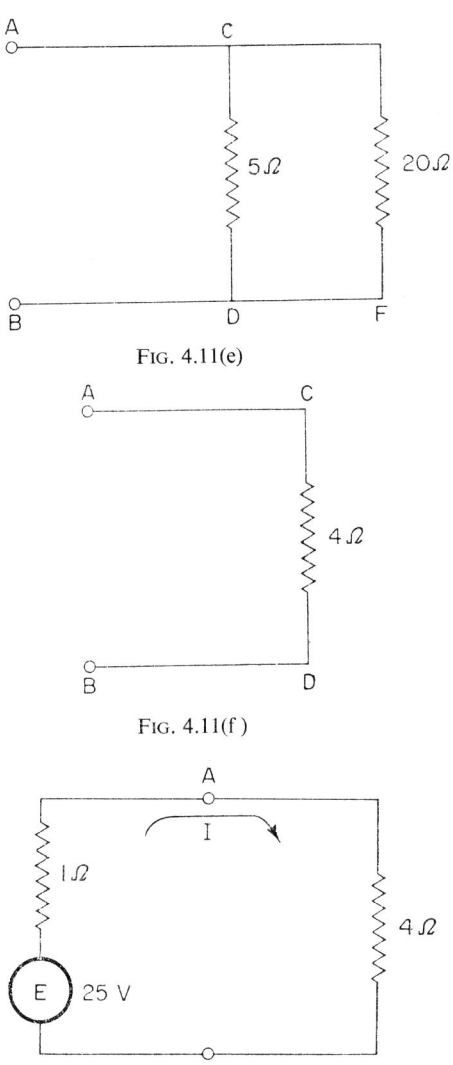

Fig. 4.11(e)

Fig. 4.11(f)

Fig. 4.11(g)

Fig. 4.11(e) ∴ Equivalent resistance from C–F

$$= 17 + 3 = 20 \ \Omega.$$

Fig. 4.11(f)   Equivalent resistance from C–D

$$= \frac{5 \times 20}{5 + 20} = \frac{100}{25} = 4 \ \Omega.$$

∴ Total network equivalent $= 4 \ \Omega$.

Now if a resistor of value $4 \ \Omega$ were connected in place of the load across A–B the e.m.f. E of 25 V would have to drive current I through its own internal resistance of $1 \ \Omega$ in series with the load of $4 \ \Omega$, Fig. 4.11(g).

$$\therefore \quad I = \frac{25}{1 + 4} = 5 \ \text{A}.$$

To find the current in each branch we use the appropriate distribution factors calculated from the expressions $[R_1/(R_1 + R_2)]$ and $[R_2/(R_1 + R_2)]$ given earlier in the chapter.

Hence from Fig. 4.11(e)

$$I_{CD} = 5 \times \frac{20}{5 + 20} = 4 \ \text{A}$$

$$I_{CF} = 5 \times \frac{5}{5 + 20} = 1 \ \text{A}$$

from Fig. 4.11(c)

$$I_{EF} = 1 \times \frac{12}{4 + 12} = \tfrac{3}{4} \ \text{A}$$

$$I_{EH} = 1 \times \frac{4}{4 + 12} = \tfrac{1}{4} \ \text{A}$$

from Fig. 4.11(a)

$$I_{JK} = \tfrac{1}{4} \times \frac{6}{3 + 6} = \tfrac{1}{6} \ \text{A}$$

$$I_{LM} = \tfrac{1}{4} \times \frac{3}{3 + 6} = \tfrac{1}{12} \ \text{A}$$

Lastly, to get the p.d. across each resistor we multiply the current through it by its ohmic value.

This gives:

$$V_{CD} = 4 \times 5 = 20 \text{ V}$$

$$V_{CE} = 1 \times 17 = 17 \text{ V}$$

$$V_{EF} = \tfrac{3}{4} \times 4 = 3 \text{ V}$$

$$V_{EG} = \tfrac{1}{4} \times 10 = 2.5 \text{ V}$$

$$V_{JK} = \tfrac{1}{6} \times 3 = \tfrac{1}{2} \text{ V}$$

$$V_{LM} = \tfrac{1}{12} \times 6 = \tfrac{1}{2} \text{ V}$$

In a practical problem it is not usually necessary to solve the network completely for every branch; very often the potential and current are required for one branch only and hence the arithmetic is much reduced.

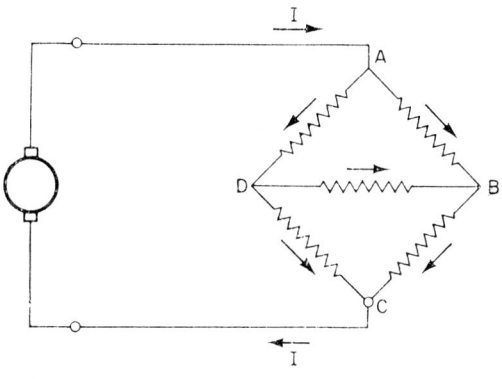

FIG. 4.12

Some networks cannot be solved by the above method because the current divided at one point does not reunite at another. A very common example of this is the network shown in Fig. 4.12 and called a Wheatstone bridge network when it is used for the determination of the value of an unknown resistor as will be explained later in the chapter.

It can be seen that the two parts of the current division at A do not unite again at another point since they are modified by the current in DB before the junction at C.

This type of network can be solved by the application of two rules called Kirchhoff's Laws which may be formulated as follows:

(1) The total current entering a junction is equal to the total current leaving it.

(2) The algebraic sum of the potential differences round any closed circuit in the network must be equal to the algebraic sum of the e.m.f's round that circuit.

Rule (1) is self-evident if it is assumed that there can be no accumulation of electricity at any point.

For Rule (2) a clear distinction must be made between potential differences and e.m.f's. The p.d. is the result of current flowing through a resistor and is equal to the product of current and resistance whilst the e.m.f. is the result of generation of electricity by one means or another which may be chemical in the case of a primary cell, magneto-mechanical as in the case of a dynamo, purely mechanical as in the case of a piezo-electric source or photoelectric as in the case of a selenium cell.

Application of Kirchhoff's Laws will always provide enough simultaneous equations to enable the network to be solved but the algebra involved is often tedious and there is the possibility of error due to a slip in the working.

A much simpler solution of networks of this type can usually be achieved by the use of two very useful transformations which follow from Kirchhoff's Laws and which should be memorized. By the application of these transformations a network can be constructed such that its voltage–current characteristic is the same as the given network but its form is so much simpler that its solution can be written down by inspection. Thus the current through any part of the original network can be obtained easily and the remainder of the network solved.

The first transformation is known as the star-delta (or conversely the delta-star) transformation. Figure 4.13(a) shows three unequal resistors, R_a, R_b and R_c connected to terminals A, B and C at one end and together at their other ends. Connexions of this type are known as *star* connexions. Figure 4.13(b) shows three different resistors R_1, R_2 and R_3 connected between terminals A, B and C in a closed circuit. This type of connexion is known as a *delta* connexion because of its resemblance to the

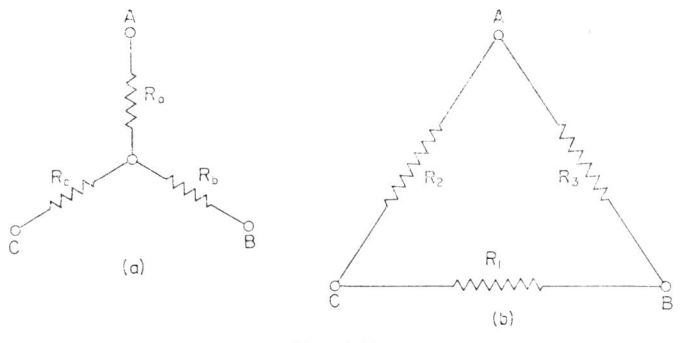

FIG. 4.13

Greek letter of that name. Alternative names for the star connexion are "wye" or "gamma" and the delta connexion is sometimes called a "mesh" connexion. Not only are single components connected in these ways but also groups of components, transformers and machines.

Now it will often be the case that in a given network difficulty in solution is due to a part of the network forming a delta (or a star) connexion between three points and in many instances if the existing connexion could be changed into an equivalent star (or delta) then the solution of the network would be much facilitated.

Figure 4.14(a) shows a complicated network of this type. By two delta-star transformations this network can be made simple.

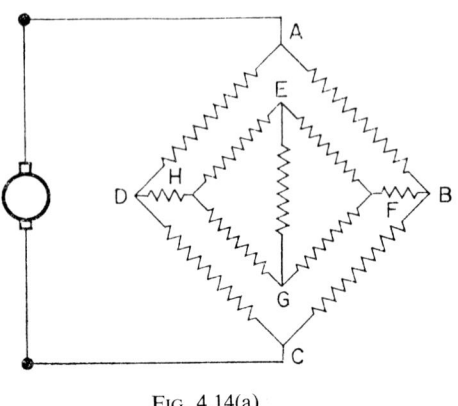

FIG. 4.14(a)

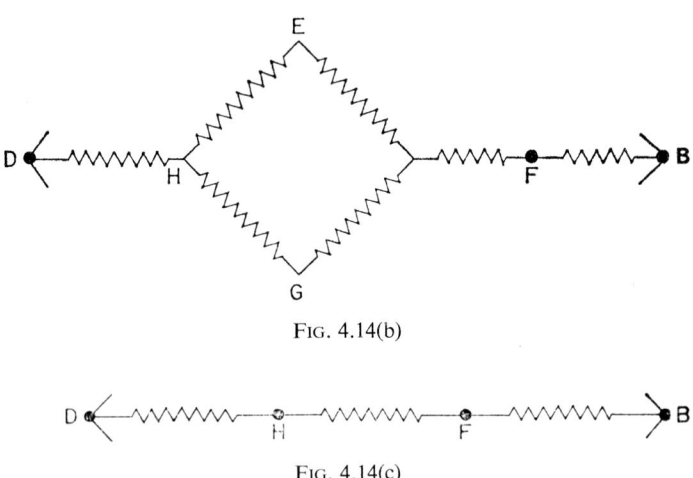

FIG. 4.14(b)

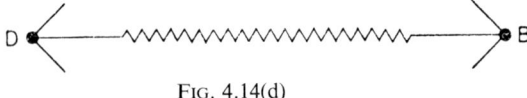

FIG. 4.14(c)

FIG. 4.14(d)

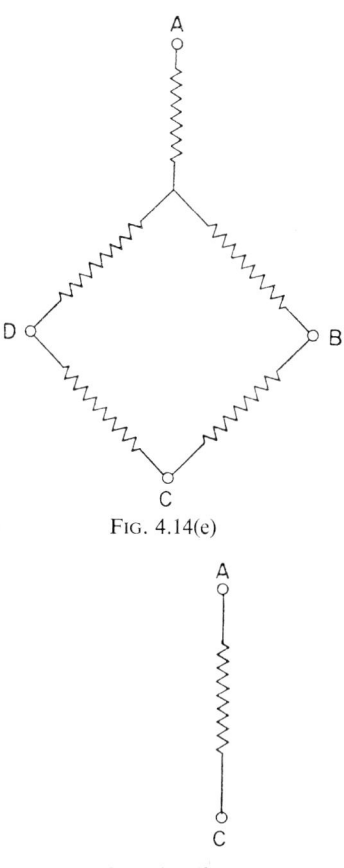

FIG. 4.14(e)

FIG. 4.14(f)

The procedure is as follows:

(1) Replace the resistors connected in delta to points E, F and G by a star as shown in Fig. 4.14(b).

(2) Determine the effective resistance between H and F, Fig. 4.14(c).

(3) Add the resistances between D and H and F and B, Fig. 4.14(d).

(4) Transform the delta ABD, Fig. 4.14(e).

(5) Determine the effective resistance between A and C, Fig. 4.14(f).

Hence the current flowing from the battery can be found and its division between the different branches determined by working backwards through stages (5) to (1).

It only remains therefore to obtain expressions for R_1, R_2 and R_3 of Fig. 4.13(b), in terms of R_a, R_b and R_c of Fig. 4.13(a) and vice versa to enable the network to be worked out numerically.

Delta to Star Transformation

Referring to Fig. 4.13.

$$R_{AB} = \frac{R_3(R_1 + R_2)}{R_1 + R_2 + R_3} \qquad \text{Fig. 4.13(b)}$$

$$R_{AB} = R_a + R_b \qquad \text{Fig. 4.13(a)}$$

$$\therefore \quad R_a + R_b = \frac{R_1R_3 + R_2R_3}{R_1 + R_2 + R_3}$$

similarly

$$R_b + R_c = \frac{R_1R_2 + R_1R_3}{R_1 + R_2 + R_3}$$

$$R_a + R_c = \frac{R_1R_2 + R_2R_3}{R_1 + R_2 + R_3}$$

from these

$$R_a - R_c = \frac{R_2R_3 - R_1R_2}{R_1 + R_2 + R_3}$$

and

$$R_a = \frac{R_2R_3}{R_1 + R_2 + R_3}$$

$$R_b = \frac{R_1R_3}{R_1 + R_2 + R_3}$$

$$R_c = \frac{R_1R_2}{R_1 + R_2 + R_3}$$

In words, the star resistance connected to a point is the product of the two delta resistances connected to that point divided by the sum of all three delta resistances.

Star to Delta Transformation

Referring to Fig. 4.13.

From the above
$$\frac{R_a}{R_b} = \frac{R_2}{R_1}$$

$$\therefore \quad R_2 = \frac{R_1 R_a}{R_b}$$

Similarly
$$\frac{R_a}{R_c} = \frac{R_3}{R_1}$$

$$\therefore \quad R_3 = \frac{R_1 R_a}{R_c}$$

Substituting for R_2 and R_3 we have

$$R_1 = R_b + R_c + \frac{R_b R_c}{R_a}$$

Similarly
$$R_2 = R_a + R_c + \frac{R_a R_c}{R_b}$$

and
$$R_3 = R_a + R_b + \frac{R_a R_b}{R_c}$$

In words, the delta resistance between two points is the sum of the star resistances connected to the points plus the product of those resistances divided by the resistance connected to the third point.

The second transformation is known as Thévénin's theorem. It enables a complete network to be represented by a single resistor in series with a single source of e.m.f. however many resistors and sources there may be in the original network. Since a common requirement is the determination of current in a single branch of a network, it is usual to state the theorem in the following way.

THÉVÉNIN'S THEOREM. *The current in any branch of a network is the same as if the branch were connected to a source of e.m.f. E_T and internal resistance R_T where, with the branch removed, E_T is the potential difference appearing across the points where the branch was connected and R_T is the effective resistance, between these two points, of the remainder of the network when all independent applied e.m.f's have been short-circuited.*

The current is given by $I = E_T/(R_T + R_B)$ where R_B is the resistance of the branch removed for calculation of E_T and R_T and replaced afterwards across the equivalent circuit as shown in Fig. 4.15.

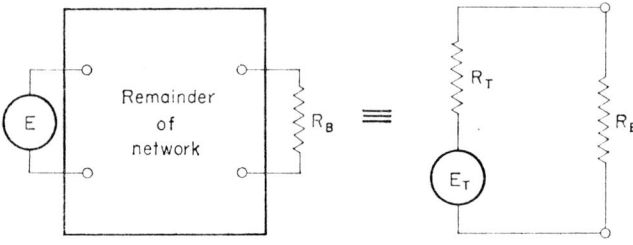

FIG. 4.15

Thévénin's theorem may be applied to any network where there is a linear relation between voltage and current for each component composing it and it may also be applied to alternating current networks.

A very common use of the Thévénin conversion is the shunted battery which can be represented diagrammatically by the circuit in Fig. 4.16. Here it will obviously be simpler to calculate the

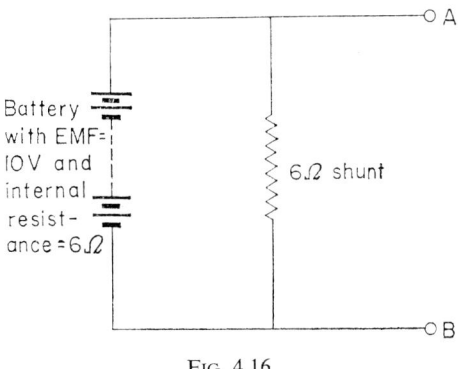

FIG. 4.16

current taken by any load connected between the terminals A and B if the supply consists of an e.m.f. and a series resistance only. The conversion is shown in stages in Fig. 4.17. V_T is the voltage appearing between A and B with no load connected and is $[6/(6 + 6)] 10 = 5$ V regarding the two 6 Ω resistors as making a potential divider, Fig. 4.17(c). R_T is the effective resistance between A and B when the source of e.m.f. is short circuited. This is 6 Ω in parallel with 6 Ω which equals 3 Ω, Fig. 4.17(d). Thus the equivalent Thévénin source is an e.m.f. of 5 V in series with a 3 Ω resistor, Fig. 4.17(c).

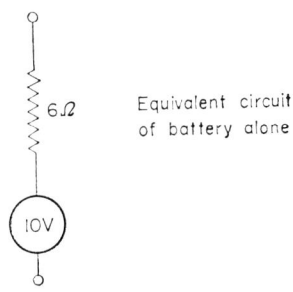

Equivalent circuit of battery alone

FIG. 4.17(a)

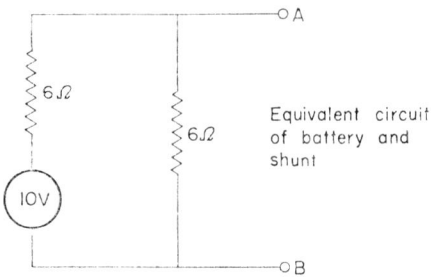

Equivalent circuit
of battery and
shunt

Fig. 4.17(b)

Circuit (b) redrawn

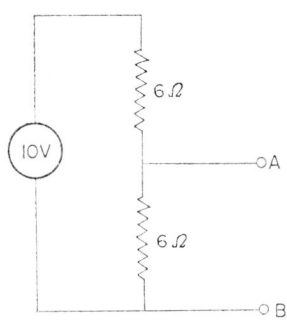

Fig. 4.17(c)

E.M.F. in
circuit (c)
now short
circuited

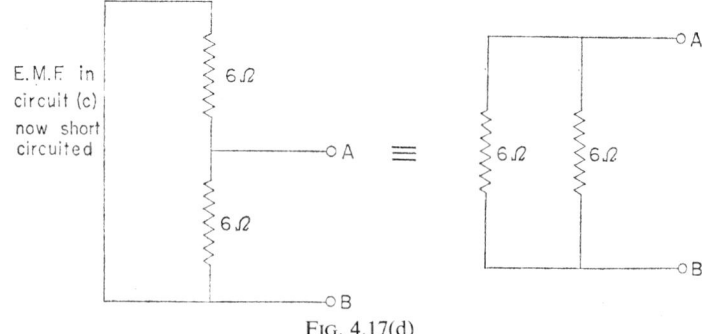

Fig. 4.17(d)

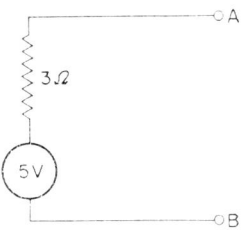

FIG. 4.17(e)

We shall find the current I in a typical Wheatstone bridge circuit, Fig. 4.18(a), by both transformation methods. First by the delta-star conversion we have:

(1) Convert delta BCD to star where

$$R_b = \frac{30 \times 40}{30 + 40 + 15} = 14\cdot12\ \Omega$$

$$R_c = \frac{30 \times 15}{30 + 40 + 15} = 5\cdot29\ \Omega$$

$$R_d = \frac{40 \times 15}{30 + 40 + 15} = 7\cdot06\ \Omega$$

(2) Redraw circuit as in Fig. 4.18(b).

(3) Calculate effective resistance of parallel circuit where $(10 + 14\cdot12)$ is in parallel with $(20 + 7\cdot06)$ giving

$$R_{\text{eff}} = \frac{24\cdot12 \times 27\cdot06}{24\cdot12 + 27\cdot06} = 12\cdot77\ \Omega$$

(4) Redraw circuit as in Fig. 4.18(c).

(5) Add to give total effective resistance across source

$$= 12\cdot77 + 5\cdot29 = 18\cdot06\ \Omega$$

$\therefore$ current from source $= \dfrac{2}{18\cdot06} = 0\cdot1108\ \text{A}$

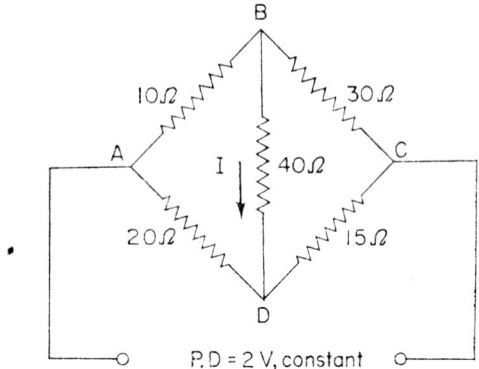

The 40 Ω resistor
represents a galvanometer

FIG. 4.18(a)

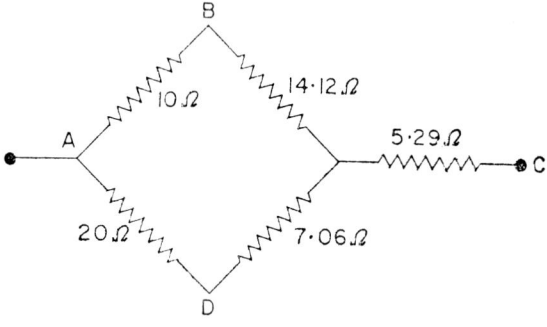

FIG. 4.18(b)

A   12·77 Ω   5·29 Ω   C

(c)

FIG. 4.18(c)

$\therefore$ division of current in Fig. 4.18(b) is given by

$$\text{upper branch current} = 0.1108 \times \frac{27.06}{51.18} = 0.0586 \text{ A}$$

$$\text{lower branch current} = 0.1108 - 0.0586 = 0.0522 \text{ A}$$

$\therefore$ p.d. A–B $= 0.0586 \times 10 = 0.586 \text{ V}$

p.d. A–D $= 0.0522 \times 20 = 1.044 \text{ V}$

$\therefore$ p.d. B–D $= 0.458 \text{ V}$

$$\therefore I = \frac{0.458}{40} = 0.0115 \text{ A}.$$

Now by Thévénin's method we have the following:
(1) Remove 40 Ω resistor connected between B and D in Fig. 4.18(a), obtaining Fig. 4.19(a).
(2) Determine p.d. between B and D from

$$\text{p.d. A–B} = 2\left(\frac{10}{30 + 10}\right) = 0.5 \text{ V since ABC is a potential}$$

divider.

$$\text{p.d. A-D} = 2\left(\frac{20}{20 + 15}\right) = 1.143 \text{ V}$$

$$\therefore \text{p.d. B–D} = 1.143 - 0.5 = 0.643 \text{ V} = V_T$$

(3) Short circuit the e.m.f., Fig. 4.19(b).
(4) Determine the effective resistance between B and D from the fact that AB is now in parallel with BC and AD is in parallel with CD. This gives equivalent resistance

$$\text{B–D} = \frac{10 \times 30}{10 + 30} + \frac{20 \times 15}{20 + 15} = 7.5 + 8.57 = 16.07 \ \Omega = R_T$$

(5) Draw equivalent Thévénin source, Fig. 4.19(c).
(6) Connect load (i.e. 40 Ω resistor removed initially) to

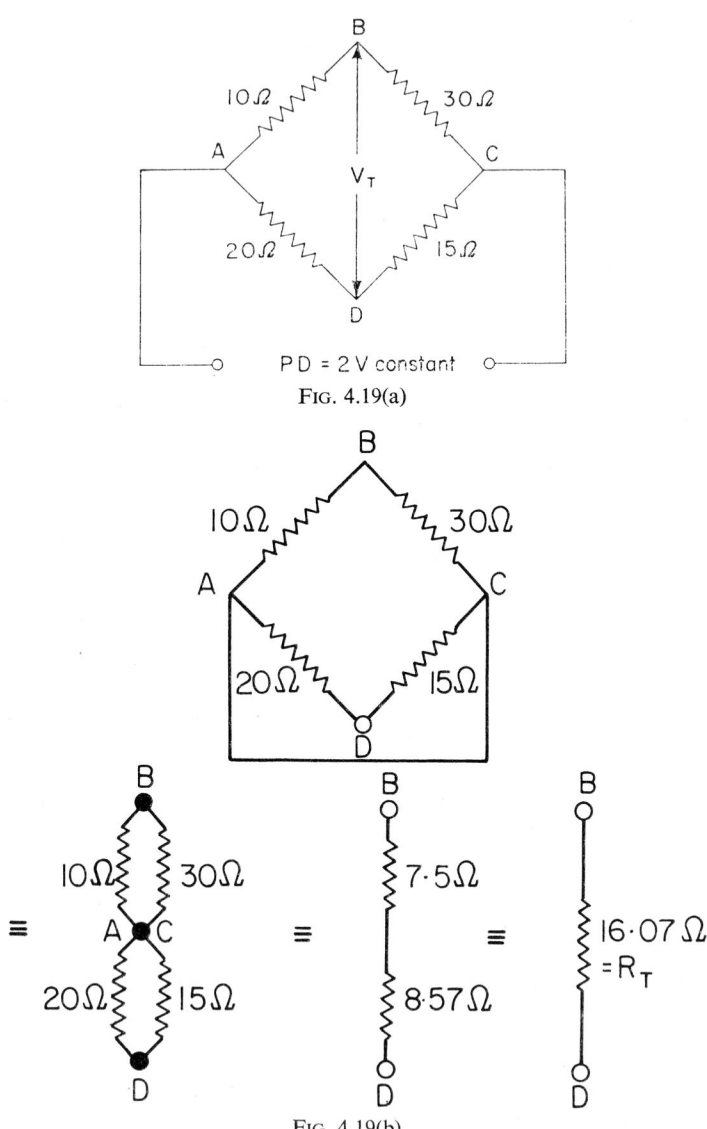

FIG. 4.19(a)

FIG. 4.19(b)

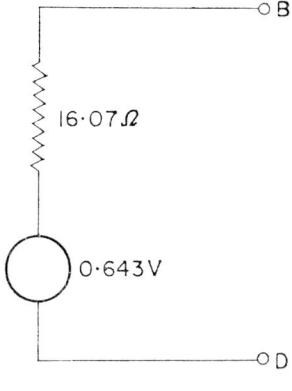

Fig. 4.19(c)

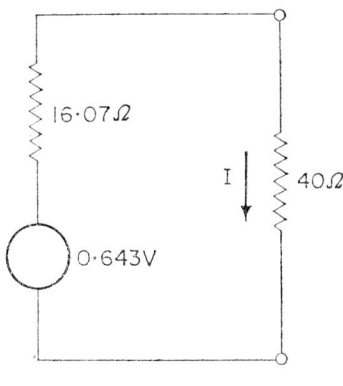

Fig. 4.19(d)

Thévenin source, Fig. 4.19(d) and calculate current flowing as follows

$$I = \frac{0 \cdot 643}{16 \cdot 07 + 40} = 0 \cdot 0115 \text{ A}$$

Which method will be shorter in a particular case depends on the type of circuit. Experience and practice in using the methods will often indicate the most appropriate one in advance.

The Wheatstone bridge circuit which has been solved above is very frequently used for the determination of the value of an unknown resistor. When so used it is usually arranged as shown in Fig. 4.20. Here arms AB and BC are known as ratio arms where the ratio of their ohmic values is known, arm AD is a calibrated variable resistor the value of which can easily be read from a scale and arm DC is the unknown resistor. In use the arm AD is varied until there is no reading on the galvanometer. This signifies that there is no potential difference between points B and D, hence the potential from B to C equals that from D to C. Since there is no current in BD all the current I_1 through AB must go through BC and all the current I_2 through AD must go through DC.

$$V_{AB} = I_1 R_{AB} = V_{AD} = I_2 R_{AD}$$

and

$$V_{BC} = I_1 R_{BC} = V_{DC} = I_2 R_{DC}$$

also

$$\frac{R_{AB}}{R_{BC}} = S, \text{ a known ratio.}$$

$$\therefore \quad \frac{V_{AD}}{V_{DC}} = \frac{R_{AB}}{R_{BC}} = S = \frac{R_{AD}}{R_{DC}}$$

$$\therefore \quad R_{DC} = \frac{R_{AD}}{S}$$

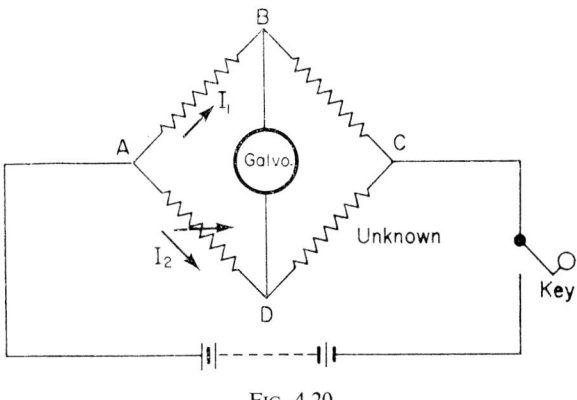

FIG. 4.20

so that the unknown resistor's value may be readily determined. If, in a particular instrument, S is permanently fixed, it is possible to make the scale associated with the adjustment of R_{AD} read the value of R_{CD} directly.

When the bridge is operated in the above manner it is said to be balanced when no current flows in the galvanometer (or other current indicating device) connected between B and D in Fig. 4.20. However it is also possible to operate the bridge in an unbalanced condition and to measure the current in BD. The advantage of doing this is rapidity in obtaining a reading without any adjustment of R_{AD}. There are, however, certain disadvantages. The relation between the current in BD and the resistance of R_{DC} is not linear—although it may be nearly so for very small differences from the value of R_{DC} which will give balance—and the range of values of R_{DC} which can be accommodated without exceeding the scale range of the galvanometer is small unless a very insensitive galvanometer is used in which case the accuracy of measurement is low. These difficulties can be overcome but only at the expense of the essential simplicity of the instrument.

Another commonly used circuit is embodied in the instrument known as a potentiometer. This is sometimes confused with the component of that name which is essentially the same in diagrammatic form and is a variable potential divider, but the component is not intended to be used as a measuring instrument and is not usually calibrated.

Basically the potentiometer, Fig. 4.21, consists of a resistor AB, which may be a straight uniform wire, and a slider C arranged to contact any point along the wire. The instrument is designed to measure the value of e.m.f. of any steady source which is connected to terminals D and E.

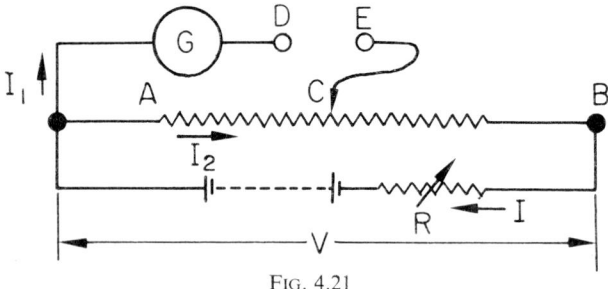

Fig. 4.21

The operation is as follows. Assuming that the source to be measured is not yet connected a current I is driven through AB by the battery and can be adjusted by the variable resistor R. For any given value of current the potential between A and C will be given by

$$V_{AC} = IR_{AC} = V\left(\frac{R_{AC}}{R_{AB}}\right)$$

By adjustment of the position of C, V_{AC} can be made to be any value between 0 and V and if the wire has a linear relation between length and resistance, the value of V_{AC} may be determined by measuring the ratio of the length AC to the length AB thus giving R_{AC}/R_{AB}. Hence a linear scale against which the position of the slider can be determined is all that is necessary for a reading of V_{AC}.

A known source is now connected to terminals D and E and, provided that the e.m.f. of this source is less than V, a position of the slider can be found so that I_1 is reduced to zero and hence no current flows in the galvanometer G. When this is so current I_2 must equal I and V_{AC} must be equal in magnitude and opposed to the e.m.f. of the known source. This gives a calibration for one point on the linear scale against which the position of the slider is read. If the slider is at A the potential between A and C is obviously zero, thus giving a second point on the scale and hence values of V_{AC} for any other points may be read directly. To further facilitate ease of operation the linear scale is usually scaled in volts, in which case it is necessary to adjust the value of I by means of the variable resistor R so that the potential V_{AC} is correct when the slider is set to the scale value equal to the known source e.m.f. and calibration using this source is being made.

Finally the unknown source is connected to D and E, the slider is moved until G reads zero, and the e.m.f. read from the scale. The accuracy of making a measurement with the instrument is obviously greatly dependent on the sensitivity of the galvanometer as it is with the Wheatstone bridge. It should also be noted that, since no current flows through the source measured when the slider has been finally adjusted, there can be no drop of potential in any internal resistance the source may have and a true reading of e.m.f. will be obtained. However, if the internal resistance of the source is very high, an extremely sensitive galvanometer will be required in order to determine just when the current through the source is actually zero.

Typical Examination Questions

1. What is the resistance looking into the terminals p and q of Fig. 1?
What is the conductance between points s and r if the terminals p and q are connected together?

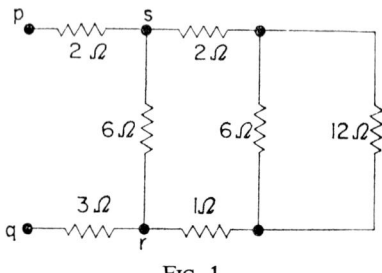

FIG. 1

Ans: 8·23 ohm, 0·51 mho.

2. Find the current output from the 10 V battery in Fig. 2.

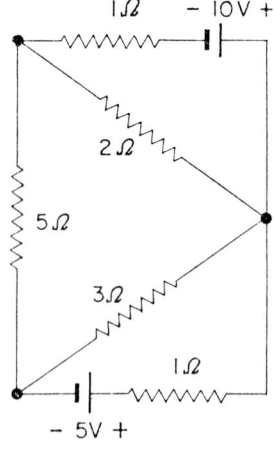

FIG. 2

Ans: 3·64 A.

3. Find the currents in the batteries in the circuit of Fig. 3.

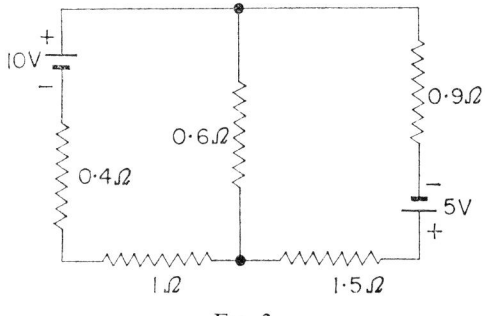

FIG. 3

Ans: 5·84 A, 2·83 A.

4. Find for the circuit of Fig. 4 the current I_0 through the terminals **TT** when R is replaced by a short circuit; the p.d. V_0 across **TT** when R is removed; and the resistance R_0 between the terminals **TT** when R is removed. Components are in ohms.

Verify that $R_0 = E_0/I_0$.

Set up the equivalent generator and find the current in R.

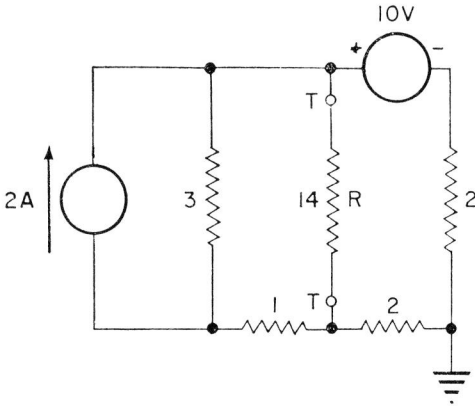

FIG. 4

Ans: 4 A, 8 V, 2 Ω, ½ A.

5. Find, in the circuit of Fig. 5, the potentials of the points a, b, c and d relative to earth.

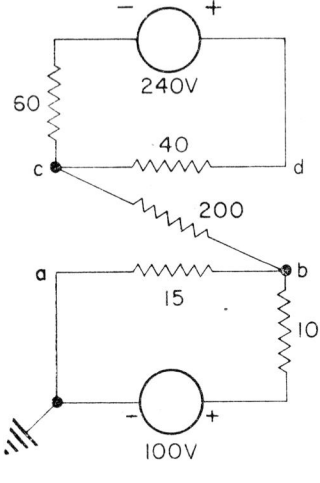

Fig. 5

Ans: a, 0; b, c, 60 V; d, 156 V.

6. A certain d.c. power supply has output p.d.'s of 600 V and 650 V when the output currents are 0·4 A and 0·2 A respectively. What simple arrangements (a) of an ideal voltage generator in series with a resistance and (b) of an ideal current generator in parallel with a conductance will give the same relation between output p.d. and current.

Ans: (a) 700 V, 250 Ω; (b) 2·8 A, 0·004 mho.

7. In the circuit of Fig. 6 A_1 and A_2 are ammeters and E_1 and E_2 are d.c. generators of constant resistance but variable voltage. An unknown linear network connects the terminals PQ to ST.

When $E_1 = 4$ V and $E_2 = 0$, the ammeter readings are $A_1 = 2$ A, $A_2 = 1$ A. What are the two possible readings of A_1 when $E_1 = 8$ V and $E_2 = 6$ V?

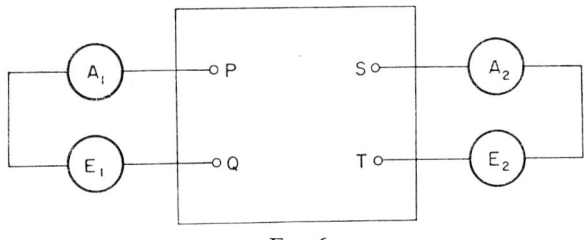

FIG. 6

Ans: 2·5 A or 5·5 A.

8. Transform the delta of Fig. 7 into its equivalent star.

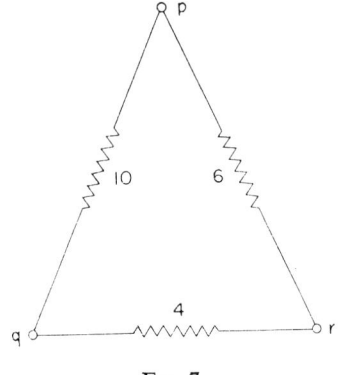

FIG. 7

Ans: Between p and star point s, 3 Ω; q and s, 2 Ω; r and s, 1·2 Ω.

9. Make use of delta-star transformations to find the resistance between p and r, and between p and s in Fig. 8.

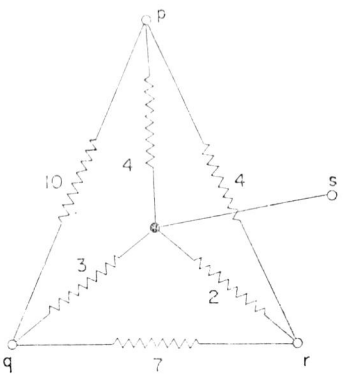

FIG. 8

Ans: 2·06 Ω, 2·02 Ω.

CHAPTER 5

Capacitive Circuits

AN IDEAL capacitor, being an electrical energy storage component rather than one which dissipates energy like a resistor, cannot, unless it is infinite in size, carry a permanent steady current nor does it have any losses. A practical capacitor has losses and also some inductance but for the purposes of the following we shall be considering ideal components.

Consider a capacitor which contains no stored electrical energy, i.e. which is uncharged. Then if electrons are taken from one plate of the capacitor and transferred by means of an external generator to the opposite plate, the capacitor becomes charged and stores electrical energy in much the same way as a spring might be compressed to store mechanical energy.

If the value of the capacitor is C farads, the quantity of electricity transferred is Q coulombs and the potential difference appearing between the terminals of the capacitor is V volts then

$$Q = VC$$

We also have from the definition of the coulomb that for a steady current

$$Q = It$$

where t is the time in seconds.

From these fundamental relationships we can determine the behaviour of a capacitor under all conditions.

The basic circuits containing more than one capacitor are the simple parallel and the simple series circuits as shown in Fig. 5.1(a) and (b).

Suppose for the circuit in Fig. 5.1(a) that the capacitors are initially uncharged. Next let charge Q_1 be transferred for C_1 and Q_2 for C_2. Because the capacitors are connected in parallel the potential difference, V, between their terminals must be the same hence, if the effective capacitance of the circuit is C, we have

$$C = \frac{Q}{V} = \frac{Q_1 + Q_2}{V} = \frac{Q_1}{V} + \frac{Q_2}{V} = C_1 + C_2$$

and this would be true for any number of capacitances in parallel.

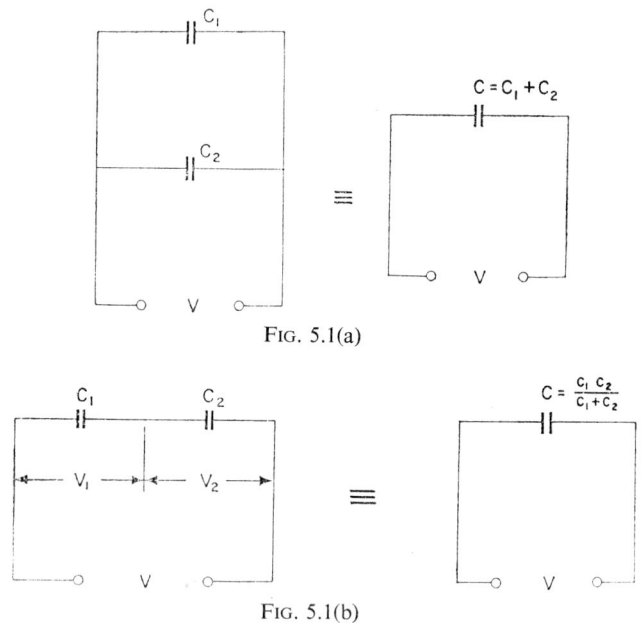

FIG. 5.1(a)

FIG. 5.1(b)

Now for the circuit in Fig. 5.1(b) assume as before that the capacitors are initially uncharged and that charges Q_1 and Q_2 are transferred in C_1 and C_2 respectively. The transfer of charge constitutes a current. Let this current, I, be held constant by suitable adjustment of the generator. It is clear that, because the capacitors are in series the current must be the same for each. Thus we have after a given time T seconds

$$Q_1 = IT \quad \text{and} \quad Q_2 = IT$$

$$\therefore \quad Q_1 = Q_2 = Q$$

but
$$V_1 = \frac{Q_1}{C_1} \quad \text{and} \quad V_2 = \frac{Q_2}{C_2}$$

$$\therefore \quad V_1 + V_2 = V = \frac{Q}{C_1} + \frac{Q}{C_2} = Q\left(\frac{1}{C_1} + \frac{1}{C_2}\right) = \frac{Q}{C}$$

$$\therefore \quad \text{equivalent capacitance } C = \frac{1}{(1/C_1 + 1/C_2)} = \frac{C_1 C_2}{C_1 + C_2}$$

This is similar in form to the equation derived in Ch. 4 for resistances in *parallel*. These expressions can readily be extended to several capacitors in parallel or series. Thus if there are capacitors C_1, C_2, ..., C_n all in parallel then the total equivalent capacitance is $C_1 + C_2 + \ldots C_n$ and if the same capacitors are all in series the equivalent capacitance is given by

$$\frac{1}{C} = \frac{1}{C_1} + \frac{1}{C_2} + \cdots + \frac{1}{C_n}.$$

More complicated networks can, of course, consist of various combinations of parallel and series connected units exactly as for the resistors and the same methods may be employed in obtaining an equivalent value for the whole arrangement. Such a network is shown in Fig. 5.2 and its solution may be obtained as follows.

(1) $C_{BD} = 3 + 2 = 5 \ \mu F$

$$\therefore \quad C_{AD} \text{ via } B = \frac{4 \times 5}{4 + 5} = 2 \cdot 22 \ \mu F$$

(2) C_{AD} via $C = \frac{2 \times 1}{2 + 1} = 0 \cdot 67 \ \mu F$

$$\therefore \quad \text{Total } C_{AD} = 2 \cdot 22 + 0 \cdot 67 = 2 \cdot 89 \ \mu F$$

(3) $V_{AB} + V_{BD} = 36 = V_{AC} + V_{CD}$

$$Q_{AB} = Q_{BD} = Q_1 \quad \text{and} \quad Q_{AC} = Q_{CD} = Q_2$$

$$\therefore \quad V_{AB} = \frac{Q_1}{4} \atop V_{BD} = \frac{Q_1}{5} \Bigg\} \text{ giving } 4V_{AB} = 5V_{BD}$$

$$V_{AC} = \frac{Q_2}{2} \atop V_{CD} = \frac{Q_2}{1} \Bigg\} \text{ giving } 2V_{AC} = V_{CD}$$

$$\therefore \quad V_{AB} = \tfrac{5}{9} \times 36 = 20 \text{ V}$$

$$V_{BD} = \tfrac{4}{9} \times 36 = 16 \text{ V}$$

$$V_{AC} = \tfrac{1}{3} \times 36 = 12 \text{ V}$$

$$V_{CD} = \tfrac{2}{3} \times 36 = 24 \text{ V}$$

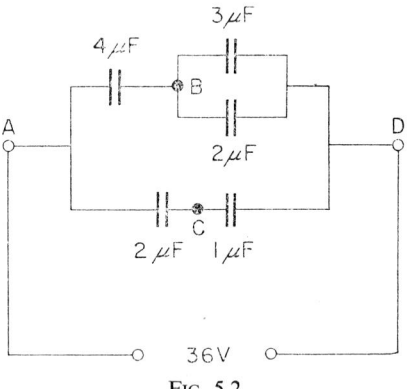

FIG. 5.2

The quantity of electricity stored in each capacitor can now be found by multiplying its capacitance by the potential difference across it.

Although capacitors are frequently used as stabilizers in circuits connected to a nominally steady source of potential, a much commoner use is in circuits which are supplied from a varying or alternating source. It is necessary, therefore, to consider the effect on charge and current of changing the p.d. applied to a capacitor. We know that at all times $Q = VC$ and we can also assume, for an ideal capacitor, that C is constant. Suppose we now change the applied p.d. by a small amount δV, then the charge will change by a small amount δQ so that

$$(Q + \delta Q) = (V + \delta V)C$$
$$\therefore \quad Q + \delta Q = CV + C\delta V$$
$$\therefore \quad \text{the extra charge } \delta Q = C\delta V$$

If this extra charge is put into the capacitor in a short time δt we have, for the *rate* at which charge and voltage change,

$$\frac{\delta Q}{\delta t} = C\frac{\delta V}{\delta t}$$

Also, if the current during the change is I

$$\delta Q = I\delta t$$

Making the small quantities infinitesimal we get

$$\frac{dQ}{dt} = C\frac{dV}{dt} = I$$

We can now see what will happen if the capacitor is connected to a source of potential varying in a sinusoidal manner as shown in Fig. 5.3.

Let the potential of the source be given at any moment by $v = V \sin \omega t$ and the value of the instantaneous current, i, must be such that

$$i = C\frac{dv}{dt}$$
$$= \frac{C\,d(V\sin\omega t)}{dt}$$
$$= \omega CV \cos \omega t$$

These results may be shown graphically as in Fig. 5.4. Any sinusoidal quantity when plotted against a linear time scale will pass through zero at regular intervals. These intervals are when $\sin \omega t = 0$ or when $\cos \omega t = 0$ which means that $\omega t = 0, \pi, 2\pi \ldots$ etc. or $\omega t = \pi/2, 3\pi/2 \ldots$ etc. respectively. Hence the time between any two successive zero values is given by

$$T = \frac{\pi}{\omega} = \frac{\pi}{2\pi f} = \frac{1}{2f} \quad \text{if} \quad f = \frac{\omega}{2\pi}$$

The symbol f is the frequency of the sinusoidal wave usually measured in c/s, and ω is sometimes called the angular frequency and is measured in rad/sec.

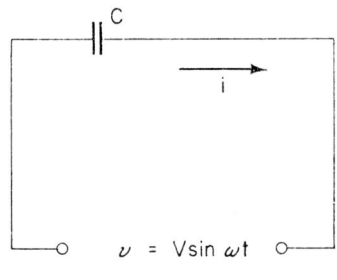

FIG. 5.3

It will be seen from Fig. 5.4(b) that the frequency of the current wave is the same as that of the voltage wave but the respective maxima are $\pi/2$ apart along the time scale. The maximum value of the voltage wave is $\pm V$, since $\sin \omega t$ can never exceed 1, and similarly the maximum value of the current wave is $\omega C V$. Thus so far as maxima are concerned

$$v_{max} \times \omega C = i_{max}$$

or

$$v_{max} = i_{max} \times \frac{1}{\omega C}$$

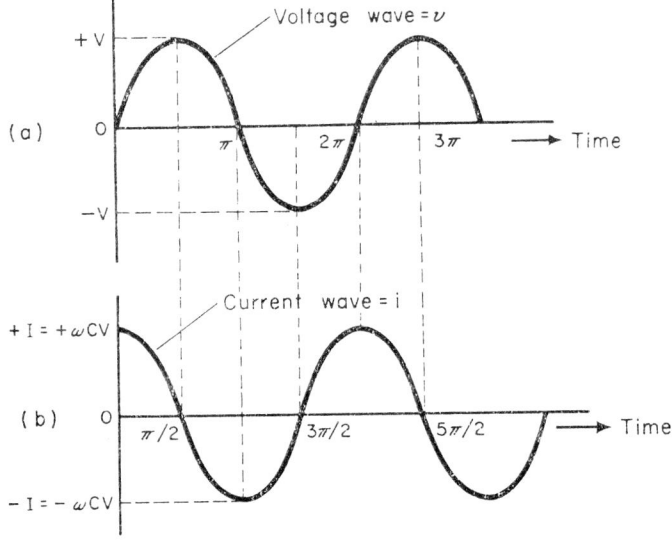

FIG. 5.4

For a particular capacitor and a particular frequency $1/\omega C$ is constant and is called the capacitive reactance of the capacitor at that frequency. The symbol for capacitive reactance is X_c.

We thus get the various relationships

$$V = IX_c$$

$$I = \frac{X_c}{V}$$

and

$$X_c = \frac{V}{I}$$

and this last expression indicates that the units of X_c are ohms because the reactance is taking the place of R in the similar expression for a pure resistance.

Looking again at Fig. 5.4(b) it will be seen that the current wave reaches its maximum *before* the voltage wave. It is therefore said to *lead* the voltage wave by $\pi/2$ or 90°. Alternatively, of course, the voltage wave may be said to *lag* the current wave by $\pi/2$.

It is not always convenient to refer to a sinusoidal quantity by its full description $V \sin \omega t$ or similarly, and in the interests of brevity a single constant value symbol is normally used. It might be thought that the peak values V or I would be sufficient but in order to achieve some link with steady voltages and currents a value which gives the same power dissipation in a resistor whether the current is alternating or direct is more suitable. Since the power in a resistor of value R carrying a steady current I is given by

$$P = I^2 R$$

we must, when the current is alternating, first sum $i^2 R$ for all values of i over one complete cycle and then take the mean value. This is shown graphically in Fig. 5.5. Then the square root of this mean value will give us the value of the alternating current

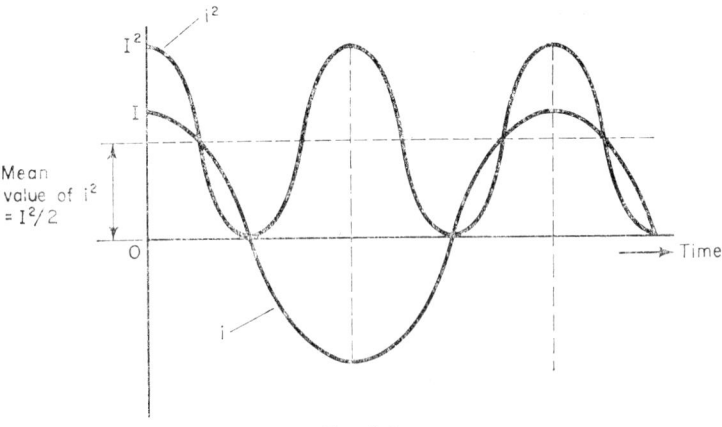

Fig. 5.5

which is equivalent in power dissipation to a steady current of the same numerical value. This current can be written I_{rms} where rms stands for *root mean square* and we can similarly refer to an alternating potential difference as V_{rms}. When there is no possibility of confusion the suffix rms is usually omitted.

A value for I_{rms} or V_{rms} can be obtained for any type of repeating waveform; for a *sinusoidal* waveform $I_{rms} = I_{peak}/\sqrt{2}$.

From the above discussion it will be seen that as long as we restrict our consideration to circuits supplied from a sinusoidal source and provided that all components in the circuits have linear relations between current and potential difference, we can solve those circuits by exactly the same methods as those used for resistive circuits carrying direct current because we can use V_{rms} and I_{rms} as constants to represent the effect of the capacitance in restricting current flow. With these provisos any of the methods given in Ch. 4 will be applicable.

Let us take as an example the circuit shown in Fig. 5.6(a).

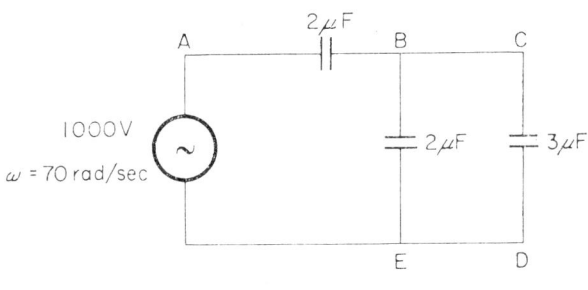

FIG. 5.6(a)

Suppose we require the rms value of the current in C_{CD}. Working in rms values and leaving out suffixes we have

(1) Capacitance of C_{BE} and C_{CD} in parallel

$$= 2\ \mu F + 3\ \mu F = 5\ \mu F$$

(2) Add C_{AB} in series giving Fig. 5.6(b).

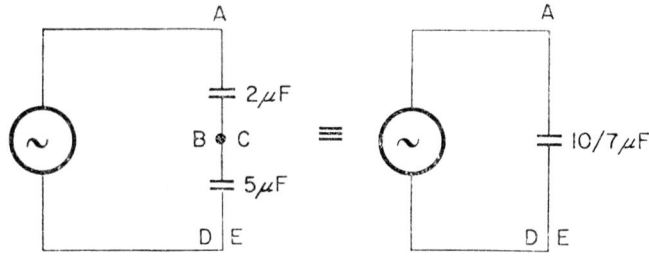

(3) Calculate equivalent capacitance across the source

$$\frac{1}{C} = \tfrac{1}{2} + \tfrac{1}{5} = \tfrac{7}{10} \quad \text{reciprocal } \mu F$$

$$\therefore \quad C = \tfrac{10}{7} \, \mu F$$

(4) Calculate reactance of C at angular frequency $\omega = 70$ rad/sec

$$X_c = \frac{1}{\omega C} = \frac{1}{70 \times \frac{10}{7} \times 10^{-6}} = \frac{10^6}{100} = 10,000 \, \Omega$$

(5) Calculate current from source

$$I = \frac{V}{X_c} = \frac{1000}{10,000} = \tfrac{1}{10} \, A$$

(6) Determine reactances X_{BE} and X_{CD}, hence find the proportions into which the current divides through C_{BE} and C_{CD}.

$$X_{BE} = \frac{1}{70 \times 2 \times 10^{-6}} = \frac{10^6}{140} \, \Omega$$

$$X_{CD} = \frac{1}{70 \times 3 \times 10^{-6}} = \frac{10^6}{210} \, \Omega$$

$$\therefore \quad \frac{I_{CD}}{I_{BE}} = \frac{140}{210}$$

$$\therefore \quad I_{CD} = I \times \frac{210}{210 + 140} = \frac{1}{10} \times \frac{210}{350} = 0.06 \, A.$$

This problem might also be solved by applying Thévénin's Theorem to the source, C_{AB} and C_{BE} regarding C_{CD} as a load.

Then as shown in Fig. 5.7 the original source and capacitance is equivalent to a source $V_T = 1000/2$ V which is the p.d. appearing across BE when C_{CD} is removed and a series capacitor 4 μF which is the effective capacitance between B and E with the source e.m.f. short-circuited.

Hence the equivalent source is 500 V with 4 μF in series. Reconnecting the 3 μF capacitor gives the circuit in Fig. 5.8. The current is given by

$$I = \frac{V}{X_c} = \frac{500}{1/(70 \times \frac{12}{7} \times 10^{-6})} = 0.06 \text{ A}$$

This method is clearly shorter than the other.

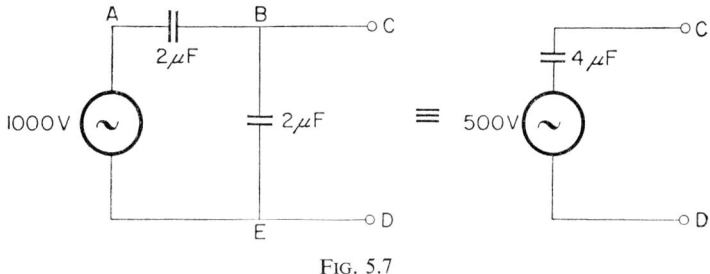

FIG. 5.7

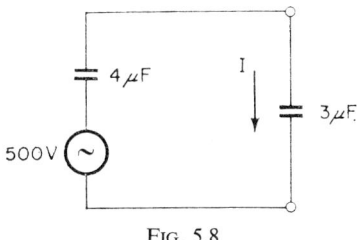

FIG. 5.8

Networks of the type shown in Fig. 5.9(a) are quite often met with. The solution of this is much facilitated if the star is transformed into an equivalent delta as shown giving a final plain delta as in Fig. 5.9(c). Note that *reactances* rather than capacitances have been given on the diagram.

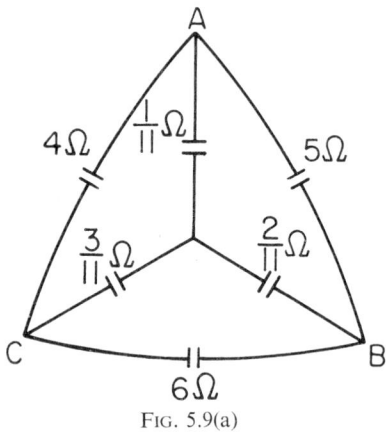

FIG. 5.9(a)

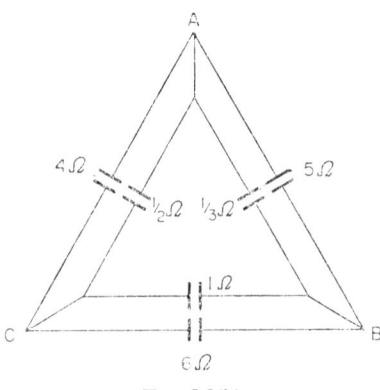

FIG. 5.9(b)

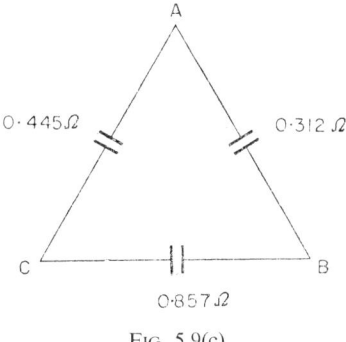

A

$0\cdot445\,\Omega$

$0\cdot312\,\Omega$

C

B

$0\cdot857\,\Omega$

FIG. 5.9(c)

The working is as follows:

(1) Convert the star to delta giving

$$X_{AB} = \frac{1}{11} + \frac{2}{11} + \left(\frac{1}{11} \times \frac{2}{11}\right)\frac{11}{3} = \tfrac{1}{3} \ \Omega$$

$$X_{AC} = \frac{1}{11} + \frac{3}{11} + \left(\frac{1}{11} \times \frac{3}{11}\right)\frac{11}{2} = \tfrac{1}{2} \ \Omega$$

$$X_{CB} = \frac{2}{11} + \frac{3}{11} + \left(\frac{2}{11} \times \frac{3}{11}\right)\frac{11}{1} = 1 \ \Omega$$

(2) Determine the total equivalent reactance between AB, AC and CB by parallel combination with the reactances already there.

$$\therefore \quad \text{Total } X_{AB} = \frac{\tfrac{1}{3} \times 5}{\tfrac{1}{3} + 5} = 0\cdot312 \ \Omega$$

$$\text{Total } X_{AC} = \frac{\tfrac{1}{2} \times 4}{\tfrac{1}{2} + 4} = 0\cdot445 \ \Omega$$

$$\text{Total } X_{CB} = \frac{1 \times 6}{1 + 6} = 0\cdot857 \ \Omega$$

If the frequency is known the capacitance of the capacitors having these reactances can readily be calculated if required.

We have now dealt with the steady state of a simple capacitive circuit having either a steady source or a sinusoidal alternating source applied. There will, however, be a transient condition for the circuit whilst the currents and p.d. are rising to their steady state values. These transient conditions are interesting and important. Some simple examples are fully dealt with in Ch. 15.

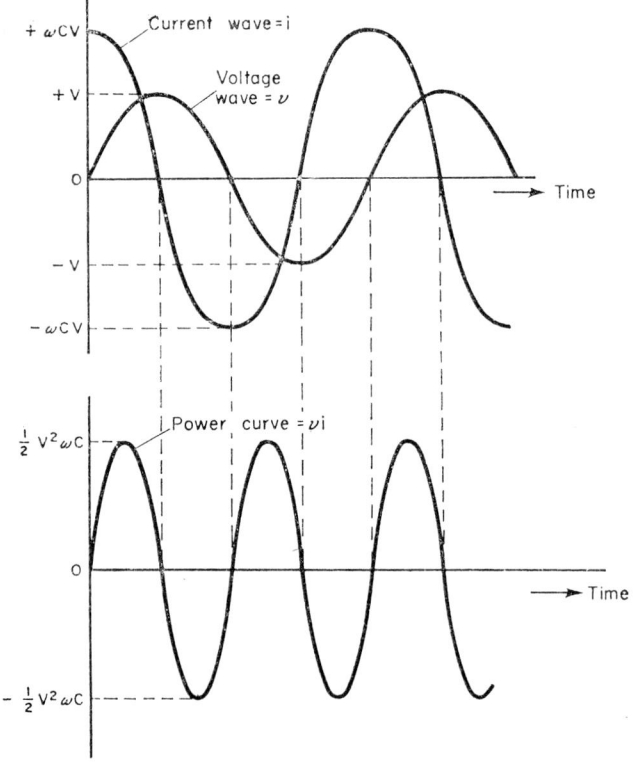

Fig. 5.10

It has been stated above that no net power is dissipated in an ideal capacitor; this can be shown as follows where the instantaneous power P is given by

$$P = vi$$

$$= V \sin \omega t \times \omega C V \cos \omega t$$

$$= V^2 \omega C \sin \omega t . \cos \omega t$$

$$= \tfrac{1}{2} V^2 \omega C \sin 2\omega t$$

The power is thus a constant $\tfrac{1}{2}V^2\omega C$ multiplied by a sinusoidally varying quantity having a frequency twice that of the voltage or current wave. The mean value of a sine wave taken over a finite number of complete cycles is zero so the net power is zero although the instantaneous power may have any value up to $\pm\tfrac{1}{2}V^2\omega C$. This means that energy is being taken into and returned from the capacitor twice every cycle of the source, but that there is no permanent accumulation of energy in it.

The same result as that given analytically above can be obtained graphically as shown in Fig. 5.10 where the power curve is obtained by multiplying the ordinates of the voltage and current curves.

Typical Examination Questions

1. Find the *change* in the capacity measured at A–B in Fig. 1 when X–Y are connected together.

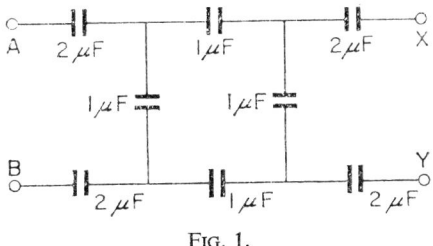

FIG. 1.

Ans: 0·012 μF.

2. A circuit consisting of a resistance R in series with an uncharged capacitance C is connected to a supply the p.d. of which is $E \sin \omega t$, where ω is a constant. If the connection is made at time t_1, show that the initial current is always $(E/R) \sin \omega t_1$ whatever value t_1 may have. Find an expression for t_1 if there is to be no subsequent transient current.

Ans: $\tan \omega t_1 = R \omega C.$

3. The phase difference of an imperfect condenser is the angle ϕ such that the vector of the current in the condenser leads the vector of the p.d. across it by $\frac{1}{2}\pi - \phi$. Show that, for constant angular frequency ω, an imperfect condenser may be represented by either

　　(a) a perfect capacitance C_s in a series with a non-inductive resistance r,

or

　　(b) a perfect capacitance C_b in parallel with a non-inductive resistance R, and that $\tan \phi = \omega r C_s = 1/\omega R C_b =$ the power factor, if ϕ is small.

Also show that $r = R/(1 + \omega^2 C_b^2 R^2)$, $R = r(1 + 1/\omega^2 C_s^2 r^2)$,

$$C_s = C_b \left(1 + \frac{1}{\omega^2/C_b^2 R^2}\right) \text{ and } C_b = C_s/(1 + \omega^2 C_s^2 r^2).$$

CHAPTER 6

Inductive Circuits

WHILST an ideal capacitor can carry no steady direct current, an ideal inductor, having no d.c. resistance, does not impede a direct current in any way. If it were connected to an ideal source having no internal resistance the current would increase without limit, the difference between the inductor thus connected and a dead short being that with the dead short the current would reach infinity immediately whilst with the inductor it would never reach infinity because there would always be a finite rate of rise of the current.

The inductor has inductance by virtue of the magnetic flux that is produced within the winding when a current flows in it. Hence any change in the value of the current will produce a change in the flux and this change will, by Lenz's Law, produce an e.m.f. in the winding of such polarity that it will oppose the potential difference producing the change in current. An inductor thus produces an effect analagous to inertia in mechanical devices. The e.m.f. produced is usually referred to as a "back" e.m.f. to indicate its opposition to the source.

If an inductor is connected by itself to a varying source of potential v the back e.m.f. must be equal and opposite to the driving potential of the source, hence we have for the latter

$$v = L\frac{di}{dt}$$

where L is the value of the inductance in henries and di/dt is the rate of change of the current through the inductor.

If we assume that the current i is sinusoidal and is expressed by

$$i = I \sin \omega t$$

then $$v = \omega L I \cos \omega t$$

and this can be shown graphically in Fig. 6.1 which is rather similar to Fig. 5.4 for the capacitor except that the current wave now *lags* on the voltage wave by $\pi/2$.

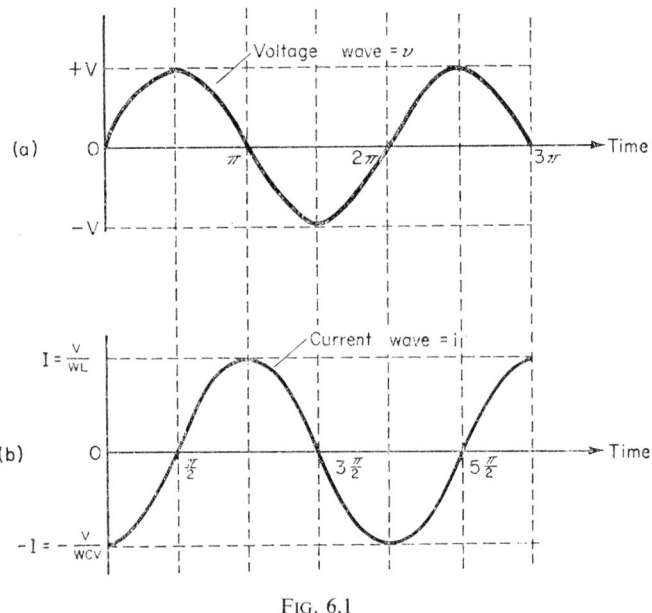

FIG. 6.1

Replacing $I \cos \omega t$ by its root mean square value we have

$$V_{rms} = \omega L I_{rms}$$

The product ωL is called the inductive reactance of the inductor and has the symbol X_L. Its unit must be ohms as it is equivalent to a voltage divided by a current.

Inductors may be connected in series or in parallel or in any combination of these. In the idealized case no inductor has any influence on any other so far as its magnetic field is concerned unless this is specifically required. In practice it is difficult to prevent part of the field of one inductor straying into the winding of another if it is adjacent to it and this frequently produces unwanted effects. There are, however, certain arrangements where by design the field of one inductor mainly or partly links another such as in a transformer. The inductors are then said to be mutually coupled.

Figure 6.2(a) shows two separate inductors L_1 and L_2 connected in series to an alternating voltage source of angular frequency ω. It is obvious that the current I will be the same for each, hence the applied p.d. is given by

$$V = X_{L_1}I + X_{L_2}I = \omega L_1 I + \omega L_2 I$$
$$= \omega I(L_1 + L_2)$$

Thus the two inductors are equivalent to one inductor having an inductance equal to the sum of the individual ones.

Figure 6.2(b) shows the parallel connection of inductors. Here it is the potential that is common to the two and the currents are given by

$$I_1 = \frac{V}{X_{L_1}} = \frac{V}{\omega L_1}$$

$$I_2 = \frac{V}{X_{L_2}} = \frac{V}{\omega L_2}$$

$$I = I_1 + I_2 = \frac{V}{\omega}\left(\frac{1}{L_1} + \frac{1}{L_2}\right) = \frac{V}{\omega} \times \frac{1}{L}$$

where
$$\frac{1}{L} = \left(\frac{1}{L_1} + \frac{1}{L_2}\right)$$

Thus for inductors in parallel, as for resistors in parallel, the reciprocal of the equivalent inductance is equal to the sum of the reciprocals of the individual inductances.

Inductors may be connected in various combinations and the solution of the circuits follows exactly on the lines described in Ch. 4 for resistors with the substitution of X_L for R. Thévénin's theorem and the star and delta transformations are equally applicable.

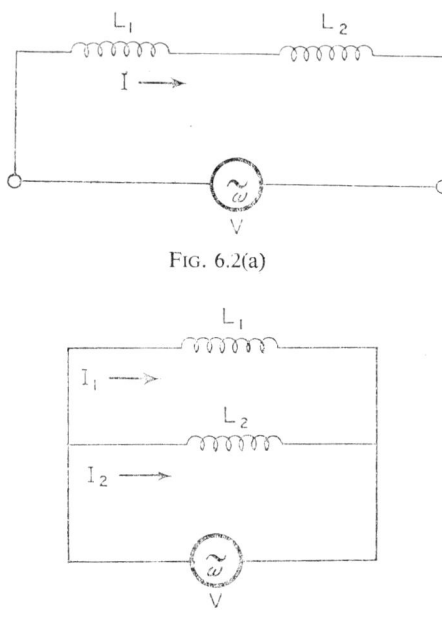

FIG. 6.2(a)

FIG. 6.2(b)

We must now consider the effect of intentional mutual coupling between inductors as this can have a profound effect on the behaviour of a circuit. The mutual inductance, M, between two coils 1 and 2 is defined by

$$M = \frac{\phi_M T_2}{I_1}$$

where ϕ_M is that portion of the whole flux produced by I_1 in coil 1 which passes also inside coil 2 which has T_2 turns. Now it will be clear that if I_1 changes there will be a change in the flux inside coil 2 and this will induce an e.m.f. in coil 2 superimposed on any e.m.f. in coil 2 which is being produced by its own current, I_2, changing. The magnitude of this e.m.f. E_M is obtained, as for the self induction e.m.f., by

$$E_M = M \frac{di}{dt}$$

where i is the instantaneous value of the current in coil 1 and M is the mutual inductance in henries.

If i is sinusoidal we have

$$E_M = \omega M I \cos \omega t$$

$$\equiv \omega M I_{rms} \quad \text{if we use rms values.}$$

It must be realized that M can have the sign $+$ or $-$ relative to any self-inductances in the circuit which means that the coils can be connected in such a way that the mutually induced e.m.f. either adds to or subtracts from the self-induced one. Which connexion is being used must be carefully ascertained before a problem can be solved numerically although it is always possible to carry a $\pm M$ through to the final expression in algebraic working.

We shall now consider the effective inductance of inductors, in series or parallel, possessing some mutual inductance.

Figure 6.3 shows two inductors in series connected in two possible ways carrying a sinusoidally varying current i.

We have for case (a)

$$\text{Coil 1} \quad V_1 = L_1 \frac{di}{dt} + M \frac{di}{dt}$$

$$\text{Coil 2} \quad V_2 = L_2 \frac{di}{dt} + M \frac{di}{dt}$$

$$\therefore \quad (V_1 + V_2) = (L_1 + L_2 + 2M) \frac{di}{dt}$$

$$\therefore \quad L_{eff} = (L_1 + L_2 + 2M)$$

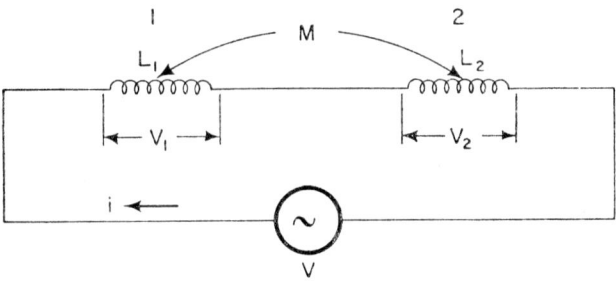

FIG. 6.3(a)

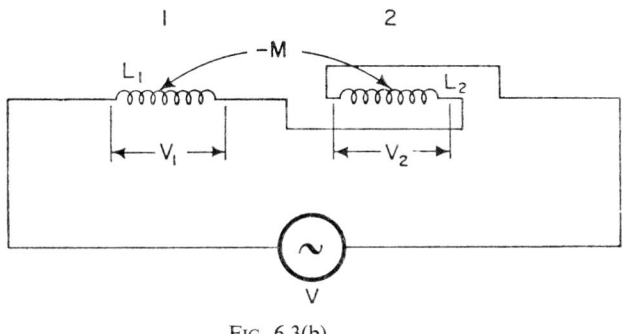

FIG. 6.3(b)

for case (b)

Coil 1 $\quad V_1 = L_1 \dfrac{di}{dt} - M \dfrac{di}{dt}$

Coil 2 $\quad V_2 = L_2 \dfrac{di}{dt} - M \dfrac{di}{dt}$

$\therefore \quad (V_1 + V_2) = (L_1 + L_2 - 2M) \dfrac{di}{dt}$

$\therefore \quad L_{\text{eff}} = (L_1 + L_2 - 2M)$

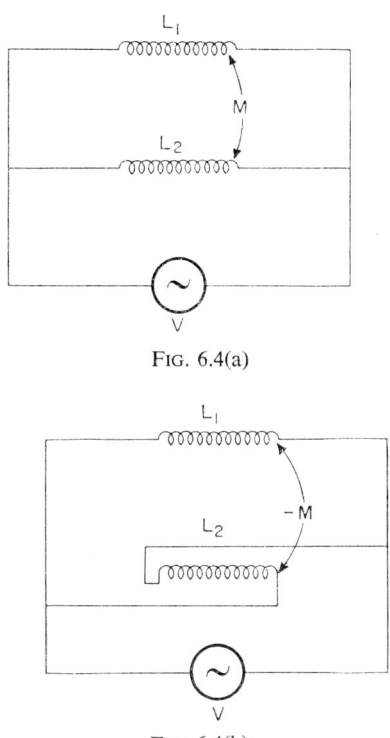

FIG. 6.4(a)

FIG. 6.4(b)

Figure 6.4 shows the two possible parallel connections. We have for case (a), assuming sinusoidal variation of V,

$$V = V_1 = V_2 = \omega L_1 I_1 + \omega M I_2 = \omega L_2 I_2 + \omega M I_1$$

$$\therefore \quad (L_1 - M)I_1 = (L_2 - M)I_2$$

$$\therefore \quad \frac{V}{\omega} = \left(L_1 + M\left(\frac{L_1 - M}{L_2 - M}\right)\right)I_1$$

$$= \left(\frac{L_1 L_2 - M^2}{L_2 - M}\right) I_1 = \left(\frac{L_1 L_2 - M^2}{L_1 - M}\right) I_2$$

$$\therefore \quad I = I_1 + I_2 = \frac{V}{\omega}\left(\frac{L_1 + L_2 - 2M}{L_1 L_2 - M^2}\right)$$

$$\therefore \quad V = I\omega\left(\frac{L_1 L_2 - M^2}{L_1 + L_2 - 2M}\right)$$

Hence

$$L_{\text{eff}} = \frac{L_1 L_2 - M^2}{(L_1 + L_2) - 2M}$$

For case (b) we replace $+M$ by $-M$ getting

$$L_{\text{eff}} = \frac{L_1 L_2 - M^2}{(L_1 + L_2) + 2M}$$

The special case of $L_1 = L_2 = L$ and $M^2 = K^2 L_1 L_2 = K^2 L^2$ is interesting. The conditions signify two identical coils where K times the flux produced by a current in one coil goes through the other coil. The value of K cannot exceed unity, for then *all* the flux from one coil would pass through the other.

Substituting in the above expressions we get for (a)

$$L_{\text{eff}} = \frac{L^2 - K^2 L^2}{2L - 2KL}$$

$$= \frac{L^2(1 - K^2)}{2L(1 - K)} = \frac{L(1 - K^2)}{2(1 - K)}$$

$$= \frac{L(1 + K)(1 - K)}{2(1 - K)} = \frac{L(1 + K)}{2}$$

If the coupling is zero and $K = 0$ then

$$L_{\text{eff}} = \frac{L}{2}$$

which we should expect from the previous analysis on p. 101 for two separate inductances in parallel. If the coupling is a maximum and $K = 1$ then

$$L_{\text{eff}} = L$$

i.e. the same as if one coil were completely removed which is a somewhat surprising result.

For case (b) we get

$$L_{eff} = \frac{L^2 - K^2L^2}{2L + 2KL}$$

$$= \frac{L^2(K^2 - 1)}{-2L(K + 1)} = \frac{L^2(1 - K^2)}{2L(1 + K)} = \frac{L(1 - K^2)}{2(1 + K)}$$

$$= \frac{L(1 + K)(1 - K)}{2(1 + K)} = \frac{L(1 - K)}{2}$$

Again if the coupling is zero and $K = 0$

$$L_{eff} = \frac{L}{2}$$

and if $K = 1$

$$L_{eff} = 0$$

We are now able to solve entirely inductive circuits such as that shown in Fig. 6.5, the analysis for which is as follows.

(1) Determine the effective reactance of the mutually coupled inductors.

$$X_{eff} = \frac{(3 \times 4) - 3^2}{(3 + 4) - (2 \times 3)}$$

$$= \frac{12 - 9}{7 - 6} = \frac{3}{1} = 3\,\Omega$$

(2) Parallel this with the 5 Ω inductor giving

$$X_{eff} = \frac{5 \times 3}{5 + 3} = \frac{15}{8} = 1\cdot875\,\Omega$$

(3) Add series inductors giving

$$\text{final } X_{eff} = 6 + 7 + 1\cdot875 = 14\cdot875\,\Omega$$

Power in an inductor can be seen to be similar to power in a capacitor, being stored every half cycle and returned to the circuit during the other half. As before the power P is given by

$$P = vi$$
$$= \omega L I \cos \omega t \times I \sin \omega t$$
$$= I^2 \omega L \cos \omega t . \sin \omega t$$
$$= \tfrac{1}{2} I^2 \omega L \sin 2\omega t$$

and the mean power is thus zero over a complete number of whole cycles.

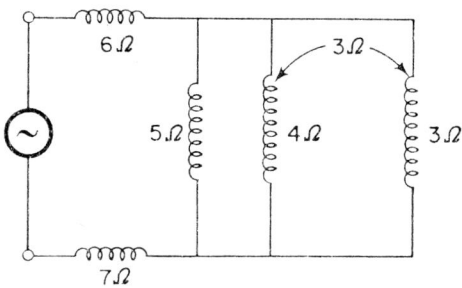

Fig. 6.5

Typical Examination Questions

1. A wooden ring of mean diameter 10 cm and of circular cross-section of diameter 1 cm carries a uniformly distributed coil of 100 turns. Calculate the value of the magnetizing force at the inner and outer edges of the wood when a current of 5 A flows in the coil. Estimate the self-inductance of the coil, assuming that the flux is uniformly distributed over the cross-section and has a value equal to that at the centre of the cross-section. Neglect the space occupied by the winding.

Ans: 1·77, 1·45 kAT/m; 3·14 μH.

2. The primary of a mutual inductance has a resistance of 20 Ω and is connected to a supply of 250 V p.d. The secondary is connected in series with an inductive load such that the resistance of the whole of the secondary

circuit is 200 Ω. At the frequency of the supply, the mutual reactance is 400 Ω, the self-reactances of the primary and secondary are each 500 Ω and that of the load is 100 Ω. Find the current taken from the supply.

Ans: 0·9 A lagging 69°.

3. A coil of resistance 2 Ω and self-inductance 0·1 H is connected to mains of p.d. $v = 283 \cos 100\pi t$ volts at the instant when $t = 0$ and the p.d. is a maximum.

Find the value of the current after a quarter of a period has elapsed, and indicate the general form of the current–time graph during a few periods.

Ans: 8·45 A.

4. Two identical coils are fixed in position relative to each other and when connected in series, their total inductance is found to be 2·0 H. The connections to one of the coils are reversed and the total series inductance is then found to be 0·4 H.

What is the coefficient of coupling between the coils?

Ans: 0·667.

5. Two coils, of negligible resistance, whose self-inductances are L_1 and L_2 respectively, have mutual inductance M and are connected in parallel. Show that the two possible values of the inductance of the combination are given by

$$L = \frac{L_1 L_2 - M^2}{L_1 + L_2 \pm 2M}.$$

CHAPTER 7

Circuits containing R, C and L

ALL PRACTICAL circuits possess some resistance, capacitance and inductance whether by design or because of imperfections in the components, so it is of considerable importance to be able to solve such networks. However any network can be broken down into simpler components which may be considered equivalent to perfect components in series or parallel so that we can represent our imperfect resistor by a combination of resistor, inductor and capacitor as shown in Fig. 7.1. Similarly we can represent our

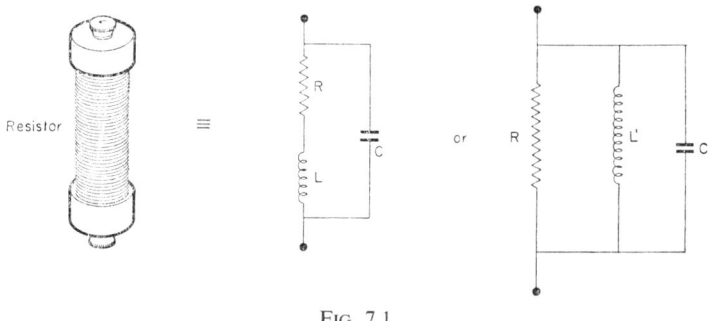

Fig. 7.1

imperfect inductor or capacitor as shown in Fig. 7.2 and Fig. 7.3. Now it will be remembered that a sinusoidal current through an inductor requires a p.d. across it which *leads* the current by 90° whilst for a capacitor the p.d. *lags* by 90°. If these p.d's are plotted on a common base as in Fig. 7.4 it can at once be seen

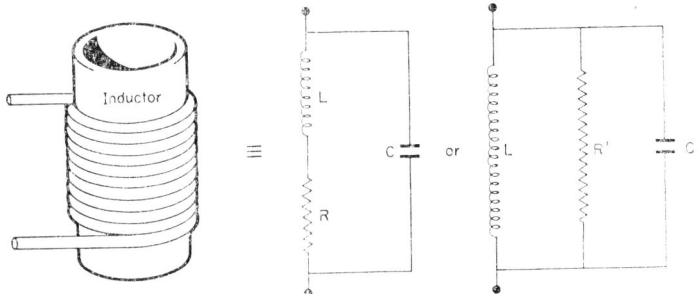

FIG. 7.2

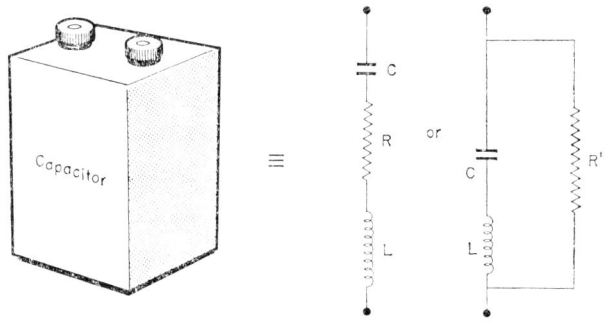

FIG. 7.3

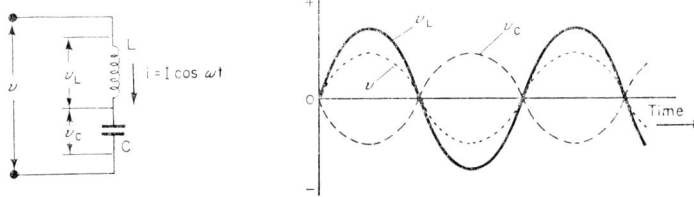

FIG. 7.4

that the effect of a capacitor is to cancel some of the effect of the inductor and vice versa if the capacitor is predominant. Therefore it will always be possible at a given frequency to represent any complex network by a simple one containing a resistor and *either* an inductor *or* a capacitor, or, in the special case when the two cancel completely, by a resistor alone.

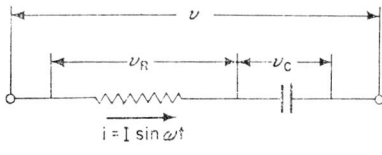

FIG. 7.5

Let us next consider some simple combinations. Figure 7.5 shows a resistor in series with a capacitor carrying a current $i = I \sin \omega t$. We know that i will require a p.d. $v_C = -V_C \cos \omega t$ across the capacitor and a p.d. $v_R = iR = RI \sin \omega t$ across the resistor. We also know that $V = I \times X_C = I/\omega C$. Thus the p.d. v across the combination as a whole is given by

$$v = v_R + v_C = RI \sin \omega t - V_C \cos \omega t$$

$$= RI \sin \omega t - \frac{I}{\omega C} \cos \omega t$$

$$= I(R \sin \omega t - X_C \cos \omega t)$$

It would be convenient if we could find an expression consisting of a constant multiplied by a single sine function to replace that in the bracket. This may be found as follows.

Multiply by Z giving

$$v = IZ \left(\frac{R}{Z} \sin \omega t - \frac{X_C}{Z} \cos \omega t \right)$$

Let X_c and R be two sides of a right-angled triangle as shown in Fig. 7.6. Then $R/Z = \cos \phi$ and $X_C/Z = \sin \phi$ and $R^2 + X^2 = Z^2$.

$$\therefore \quad Z = \sqrt{(R^2 + X_C^2)}$$

$$\therefore \quad v = IZ(\cos \phi \sin \omega t - \sin \phi \cos \omega t)$$

$$= IZ \sin(\omega t - \phi) \quad \text{where} \quad \tan \phi = X_C/R$$

The quantity Z is called the impedance of the circuit and is measured in ohms; the angle ϕ is the phase angle or angle of

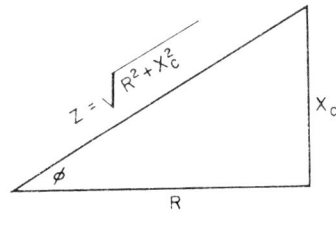

FIG. 7.6

lead or lag between the voltage wave and the current wave. This can clearly be seen in Fig. 7.7 which is the graphical solution of the above expression obtained by adding the voltage waves for the resistor and capacitor. It can be seen that the result is a wave of peak magnitude IZ and *lagging* the current wave by a phase angle ϕ.

We shall now consider the circuit shown in Fig. 7.8, where we have an inductor in series with a resistor. The current is $i = I \sin \omega t$ and the voltage v is given by

$$v = RI \sin \omega t + X_L I \cos \omega t$$

$$= IZ\left(\frac{R}{Z} \sin \omega t + \frac{X_L}{Z} \cos \omega t\right)$$

$$= IZ \sin(\omega t + \phi)$$

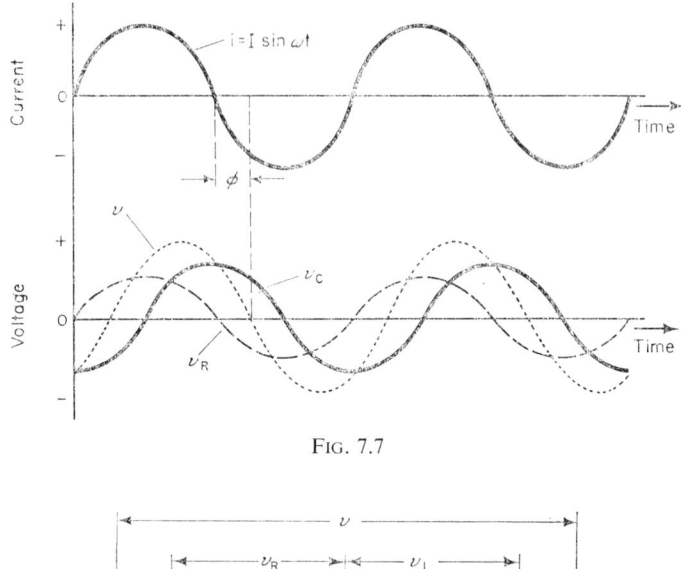

FIG. 7.7

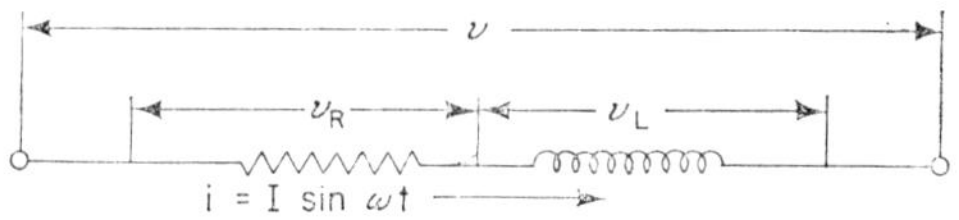

FIG. 7.8

As before, Z is the impedance of the circuit and is equal to

$$\sqrt{(R^2 + X_L^2)} \text{ ohms, and } \tan \phi = X_L/R$$

This time the graphical solution gives Fig. 7.9 where the applied voltage *leads* the current by the angle ϕ.

It is now a simple matter to combine these results and apply them to the circuit in Fig. 7.10 where all three components are in series.

The total voltage across the circuit v is now given by

$$v = RI \sin \omega t + X_L I \cos \omega t - X_C I \cos \omega t$$

$$= RI \sin \omega t + (X_L - X_C)I \cos \omega t$$

Obviously this expression is the same as the previous ones except that $(X_L - X_C)$ takes the place of X_C or X_L alone. Thus the impedance Z is now given by

$$Z = \sqrt{(R^2 + (X_L - X_C)^2)}$$

and the phase angle by

$$\tan \phi = \frac{(X_L - X_C)}{R}$$

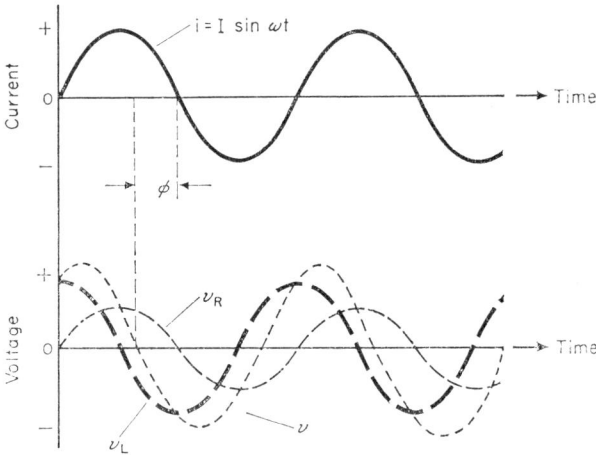

FIG. 7.9

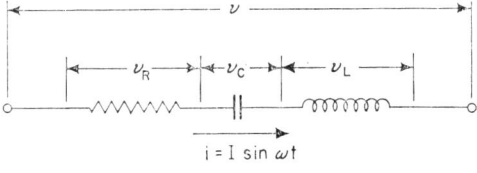

FIG. 7.10

The expression $(X_L - X_C)$ is the net reactance of the circuit and will in general obviously be inductive or capacitive depending on which of the two component reactances is the greater. Thus if X_L is greater than X_C the remainder after subtraction is inductive and the bracket is +ve. If X_C is greater than X_L the remainder is capacitive and the bracket is $-$ve.

If $X_L = X_C$ (i.e. $\omega L = 1/\omega C$) then the circuit has no net reactance, is purely resistive and Z is a minimum. The circuit is then said to be in a *resonant* state at *resonant frequency* given by $\omega = \sqrt{(1/LC)}$.

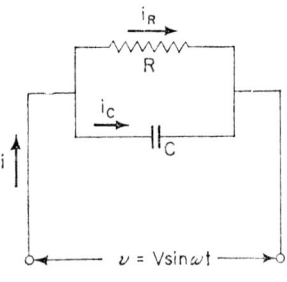

FIG. 7.11

We must next consider the parallel circuit shown in Fig. 7.11.

Here the voltage $v = V \sin \omega t$ is common to the two components so that the total current is given by

$$i = i_R + i_C$$

$$= \frac{V}{R} \sin \omega t + \frac{V}{X_C} \cos \omega t$$

$$= \frac{V}{Z} \left(\frac{Z}{R} \sin \omega t + \frac{Z}{X_C} \cos \omega t \right)$$

$$= \frac{V}{Z} \sin (wt + \phi) \text{ as before}$$

and
$$Z = \frac{1}{\sqrt{(1/R^2 + 1/X_C^2)}} \quad \text{and} \quad \tan \phi = \frac{R}{X_C}$$

Similarly the expression for total current for the circuit shown in Fig. 7.12 is

$$i = \frac{V}{Z} \sin (wt - \phi)$$

where

$$Z = \frac{1}{\sqrt{(1/R^2 + 1/X_C^2)}}$$

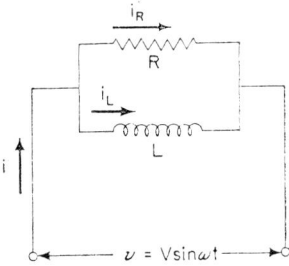

FIG. 7.12

Practical inductances and capacitances have some inherent resistance so that for an actual combination of these components we would have a circuit as shown in Fig. 7.13 for two such components in parallel. The analysis of the circuit is as follows:

$$Z_1 = \sqrt{(R_1^2 + \omega^2 L^2)} \qquad Z_2 = \sqrt{(R_2^2 + (1/\omega C)^2)}$$

$$\tan \phi_1 = \frac{\omega L}{R_1} \qquad \tan \phi_2 = \frac{1}{\omega C R_2}$$

$$\cos\phi_1 = \frac{R_1}{Z_1} \qquad\qquad \cos\phi_2 = \frac{R_2}{Z_2}$$

$$\sin\phi_1 = \frac{\omega L}{Z_1} \qquad\qquad \sin\phi_2 = \frac{1/\omega C}{Z_2}$$

$$i_1 = \frac{V}{Z_1}\sin(\omega t - \phi_1) \qquad\qquad i_2 = \frac{V}{Z_2}\sin(\omega t + \phi_2)$$

$$i = i_1 + i_2 = V\left[\left(\frac{\cos\phi_1}{Z_1} + \frac{\cos\phi_2}{Z_2}\right)\sin\omega t\right.$$

$$\left. - \left(\frac{\sin\phi_1}{Z_1} + \frac{\sin\phi_2}{Z_2}\right)\cos\omega t\right]$$

$$= V\left[\left(\frac{R_1}{Z_1^2} + \frac{R_2}{Z_2^2}\right)\sin\omega t - \left(\frac{\omega L}{Z_1^2} - \frac{1/\omega C}{Z_2^2}\right)\cos\omega t\right]$$

$$= V(A\sin\omega t - B\cos\omega t)$$

where
$$A = \left(\frac{R_1}{Z_1^2} + \frac{R_2}{Z_2^2}\right), \qquad B = \left(\frac{\omega L}{Z_1^2} - \frac{1/\omega C}{Z_2^2}\right)$$

$$\therefore \quad i = V\sqrt{(A^2 + B^2)}\left(\frac{A}{\sqrt{(A^2+B^2)}}\sin\omega t - \frac{B}{\sqrt{(A^2+B^2)}}\cos\omega t\right)$$

$$= V\sqrt{(A^2 + B^2)}\sin(\omega t - \phi) = I\sin(\omega t - \phi)$$

$$\therefore \quad I = V\sqrt{(A^2 + B^2)} \quad\text{and}\quad Z = \frac{V}{T} = \frac{1}{\sqrt{(A^2 + B^2)}}$$

and
$$\tan\phi = \frac{B}{A}$$

By a suitable variation of parameters the net reactance of this circuit may also be made zero and the circuit will thus be resonant at the appropriate frequency.

So far in all these expressions we have given the voltages and currents in their sinusoidal forms. It is however usual, as shown in the previous chapter, to use rms values which obviates writing down sine and cosine functions. It is clear from the various

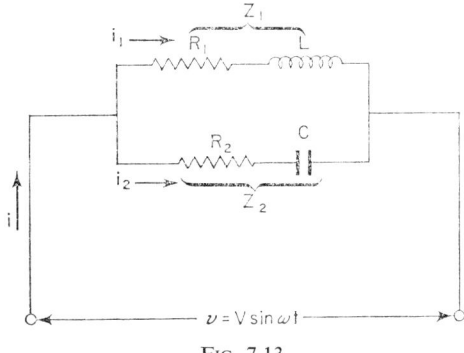

FIG. 7.13

expressions developed above that the maximum values (and hence the rms values for a sine or cosine wave) do not depend on the phase angle ϕ. Hence the rms values may be obtained from the various expressions by simply omitting the sine or cosine function in each case and dividing the peak values by $\sqrt{2}$. If the phase angle is required it can always be obtained from tan ϕ = net reactance/resistance.

Typical Examination Questions

1. An a.c. of 1 kA is transmitted by two circuits in parallel; one of these has a resistance of 2 Ω and negligible reactance, the other has a resistance of 2 Ω and a reactance of 2 Ω. Find the current in each circuit.

Ans: 633, 447 A.

2. A p.d. of 250 V a.c. is maintained across a circuit consisting of a 5 Ω resistance in series with a coil which has a resistance of 8 Ω and a reactance of 6 Ω; a 10 Ω resistance is connected as a shunt across the terminals of the coil. In the steady state, what is the p.d. across the coil?

Ans: 130 V.

3. A circuit with resistances and reactances as shown in Fig. 1 is supplied from a.c. mains of 200 V p.d. Find the p.d. across the condenser and its phase relative to the p.d. of the mains.

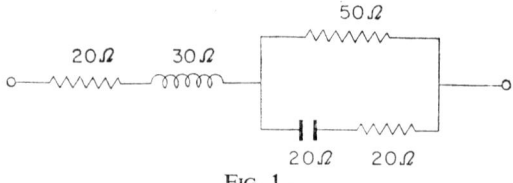

FIG. 1

Ans: 65·0 V lagging by 103°.

4. A resistance AB of 1 Ω, a reactance BC of 2 Ω, a resistance CD of 20 Ω and a capacity reactance DA of 20 Ω are joined in a closed circuit $ABCD$. An alternating p.d. of 10 V is maintained across AC. Find the p.d. between B and D.

Ans: 3·16 V.

5. A condenser of capacitance C μF, in parallel with a non-inductive resistance of R ohms, is connected to an alternator giving a constant p.d. of 150 V at various frequencies. At a frequency of 50 c/s the steady current in the joint circuit is 0·08 A; at a frequency of 90 c/s it is 0·10 A. Find C and R.

Ans: 0·85 μF; 2160 Ω.

6. Show that the balance conditions for the bridge shown in Fig. 2 is

$$\frac{C_1}{C_2} = \frac{S}{Q} = \frac{r}{R},$$

where C_2, r is the imperfect condenser to be compared with the perfect condenser C_1. Show that the phase difference of C_2 is $\tan^{-1} pRC_1$.

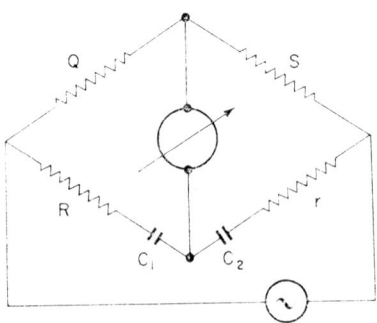

FIG. 2

CHAPTER 8

Graphical Methods

IN THE previous chapters we have considered the flow of current in various parts of a network and the resultant voltages developed across the various branches when the quantities were always sinusoidal in wave form. This wave form has a useful property that it can be derived from the projection of a point moving at a steady speed round a circle onto a diameter of that circle. Figure 8.1(a) shows such a point P_1 and the motion of the point S, which is always vertically below P_1, along the line AOB will be sinusoidal. If there were a pencil at point S and a strip of paper were to be moved at a steady speed beneath the pencil in the direction DO then a sinusoidal wave of the familiar shape and of peak value R_1 as shown in Fig. 8.1(b) would be drawn out on the paper. If now a second point P_2 is taken on another circle where the radius OP_2 is permanently behind OP_1 by an angle ϕ as shown in Fig. 8.2(a) then a second point S_2 on the line AOB will also have sinusoidal motion and a second pencil at this point would draw out another curve of the same shape as the first but displaced along the direction of motion by a distance equal to $\phi/360$ times the wavelength. The peak value of this wave will be R_2, as shown in Fig. 8.2(b) where the spacing between the waves has been exaggerated a little in the interests of clarity.

It is easy to see that any of the waves shown in the last chapter could be drawn by appropriately placed pencils moving as projections of points moving in circles having a common centre, the

121

frequency of the wave being the number of times per second the point travelled round its circle.

The line joining such a point to the centre of the circle is called a vector. The length of this line is the peak value of the wave it represents and its angular displacement from other lines gives the phase displacement of the waves, but it should be noted that these vectors are not true vectors in that current and voltage do

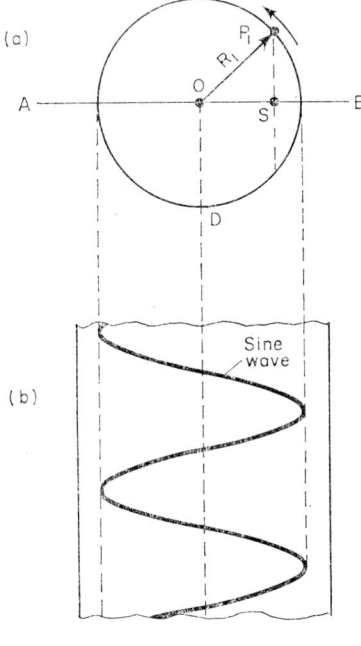

Fig. 8.1

not have a direction in space represented by the vector. There is one further property of vectors which makes them particularly suitable for the representation of currents and voltages in networks which is that the vector representing the sum or difference of two or more sinusoidal quantities of the same frequency can

very easily be obtained graphically by means of a simple geometrical construction shown in Fig. 8.3. It will be seen that the
vectorial sum of vectors OA and OB is OC which is obtained by
completing the parallelogram $ACBC$ and drawing the diagonal
OC. This vector has the correct length for the peak value of the

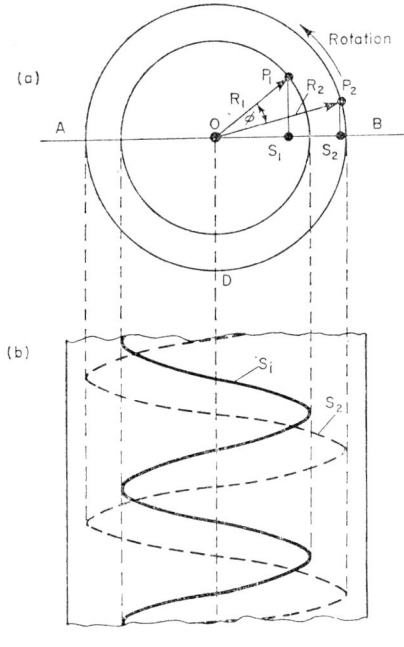

Fig. 8.2

sum of the original waves and it is at the correct angle ϕ in advance
of the vector OB and the correct angle θ behind OA. Note that
the words "advance" and "behind" depend on the direction in
which the vectors are arbitrarily arranged to rotate. If they rotate
anticlockwise as shown in Fig. 8.3 which is the conventional

direction, then the phase displacements are as stated above. If the rate of rotation is f revolutions per second then the frequency represented is f c/s and the time for one cycle is $1/f$ seconds.

Since there are 2π rad in one revolution there are $2\pi f$ rad/sec for a wave of frequency f c/s.

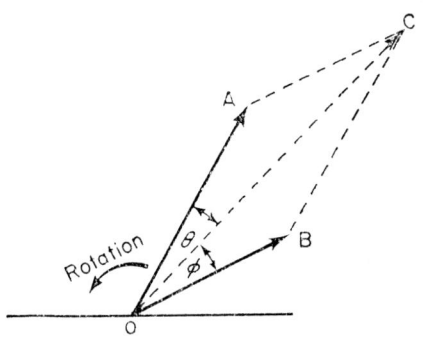

FIG. 8.3

Let us now see how the voltage and current may be represented vectorially for our three basic circuit elements.

Figure 8.4 shows (a) the vectors for a resistor, (b) the vectors for a capacitor and (c) those for an inductor. It is clear that the voltage wave *lags* the current for the capacitor and *leads* for the inductor by 90°.

Next we will combine the elements in pairs. Figure 8.5 shows a resistor in series with a capacitor. The current is common to the two components so that the vector representing this is drawn first and then a voltage vector of length IR is drawn on top of (i.e. in phase with) the current vector and a second vector of length IX_C is drawn 90° lagging on the current vector. The sum

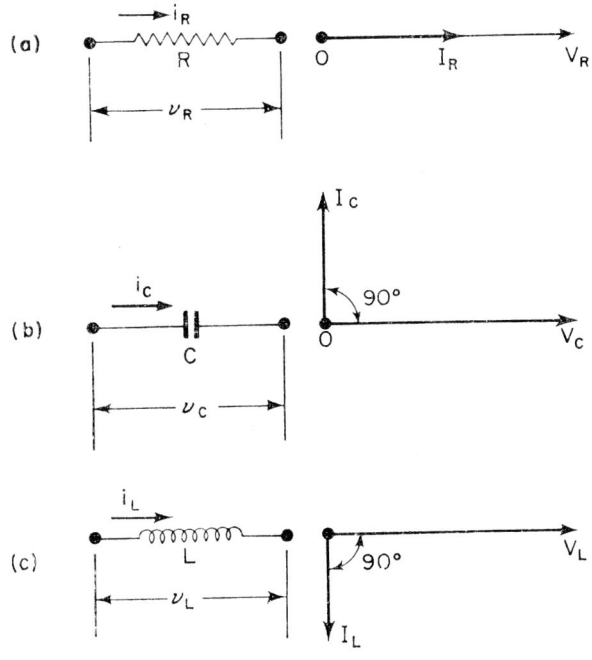

FIG. 8.4

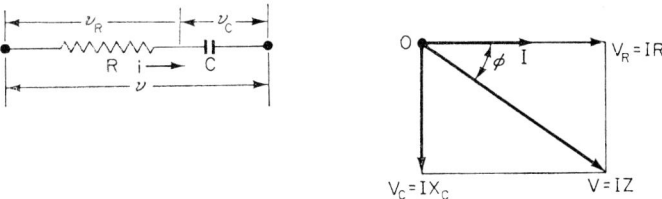

FIG. 8.5

of these two voltage vectors is obtained by completing the parallelogram to give vector V. The length of this vector is, by
Pythagoras, given by

$$V = \sqrt{(I^2R^2 + I^2X_C^2)}$$

$$= I\sqrt{(R^2 + X_C^2)}$$

$$= IZ$$

where Z is the impedance of the circuit and the phase angle ϕ is
given by

$$\tan \phi = \frac{IX_C}{IR} = \frac{X_C}{R}$$

these results, of course, being the same as those obtained earlier
by another method.

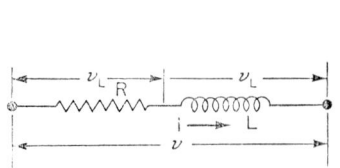
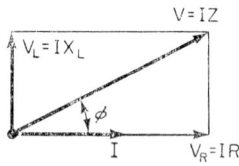

FIG. 8.6

Similarly Fig. 8.6 shows a resistor in series with an inductor
and the associated vector diagram. This is much the same as
that in Fig. 8.5 except that the voltage vector for the inductor
leads the current by 90°.

As before

$$V = I\sqrt{(R^2 + X_L^2)} = IZ$$

and $$\tan \phi = \frac{X_L}{R}$$

Figure 8.7 shows the diagrams for the parallel circuit. Here it is the voltage that is common to the circuit and hence this vector is drawn first. The current in the resistor must be in phase with the voltage and the current in the capacitor must *lead* by 90° and the total current *I leads* the voltage *V* by the angle ϕ.

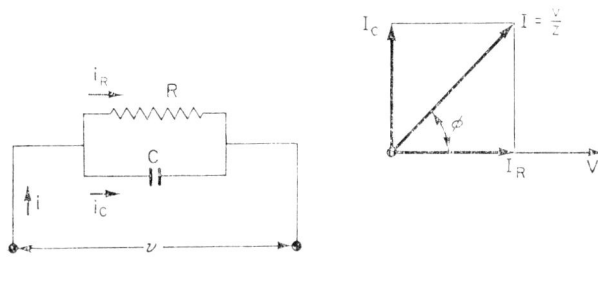

FIG. 8.7

We can now derive the diagrams for all three elements in series or in parallel.

Figure 8.8 shows the series case. As before the current is common and three voltage vectors must be drawn as follows: one in phase (for the resistor), one lagging by 90° (for the capacitor) and one leading by 90° (for the inductor). These three vectors must now be added vectorially. This is done by adding any two and then adding the resultant vector to the third. It is obviously simplest to add the V_L and V_C vectors first as they are diametrically opposed and then to add the result $V_L - V_C$ to V_R by completing the parallelogram.

Figure 8.9 shows the parallel case. Here the common vector is the voltage one and the vectors to be added are the current ones giving the final resultant current *I*.

So far we have quite correctly considered the lengths of the vectors to represent the peak values of the various waves. However since the rms value of a sine wave is a fixed proportion,

$1/\sqrt{2}$, of the peak value we might just as well represent the rms values by vectors of length equal to the peak/$\sqrt{2}$ and hence read of our final values directly in rms. For the remainder of this chapter this will be done unless otherwise stated.

In many practical instances of the application of the vectorial method it is necessary to start the vector diagram with a vector

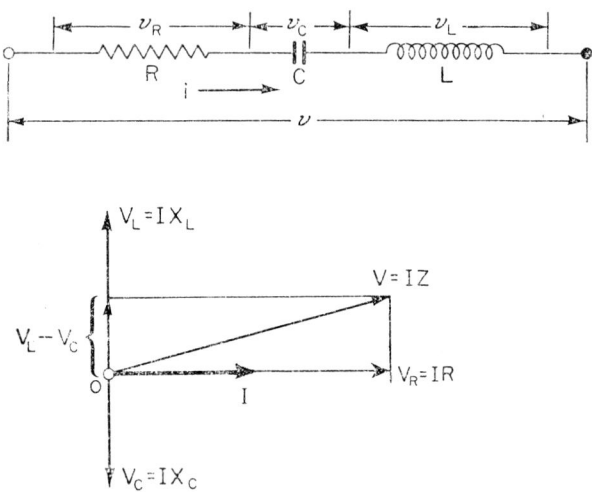

FIG. 8.8

of unknown length. For instance in Fig. 8.8 above the voltage applied to the circuit might have been known, but not the current, yet to draw the vector diagram conveniently it is necessary to start with the common parameter, the current. In these cases, bearing in mind that a vector diagram is a *scale* diagram, it is usual to assume a current of say 1 A. This is then drawn and the lengths of the remainder of the vectors calculated on this assumption. Then when the final vector has been drawn—in this case the voltage across the circuit—it will not be equal (unless the current actually happened to be 1 A) to the voltage given. It is now only necessary to rescale the whole diagram so that the voltage vector length is made equal to the given voltage and the length originally chosen to represent 1 A will now give the actual current taken.

The essence of the method, therefore, is the construction of a set of vectors all correctly located angularly and of appropriate

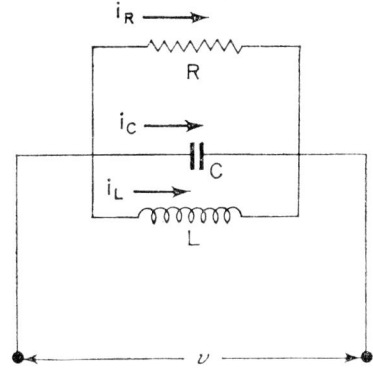

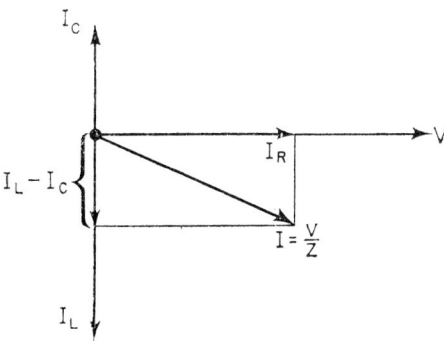

Fig. 8.9

relative length so that the voltage and current in every branch of the network is represented. The construction can be commenced at any point in the network where a potential difference or a current, or possibly both, are known and where the reactance or resistance of every branch is given. It is sometimes possible and very useful to construct the diagram so as to allow for an unknown reactance and hence to determine the value of this reactance which will fit certain given conditions, but in general circuits with unknown impedances are better solved analytically.

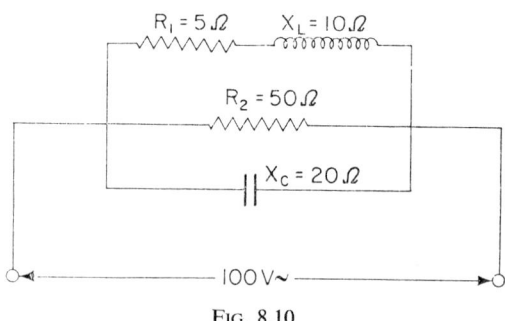

Fig. 8.10

We must now consider the solution of circuits more complicated than the simple series or parallel ones but using the solutions of these as building blocks.

Figure 8.10 shows a very common mixed circuit having a series part consisting of an inductor and a resistor all in parallel with a capacitor and resistor. This could be a representation with ideal components of an actual coil, which would have some inherent series resistance, in parallel with a real capacitor, which would have some leakage resistance in parallel. The whole is connected to a supply of p.d. 100 V.

The stages of the solution are as follows.

(1) Choose suitable scales for volts and amps consistent with the data and size of paper.

(2) Assuming a current of 1 A in the inductor, draw the vector diagram for the inductor and resistor in series making $V_L = 10 \times 1 = 10$ V and $V_R = 5 \times 1 = 5$ V. Complete the parallelogram to obtain V which on measurement is 11·2 V. The given value of V is 100 V therefore the scale of the diagram must be multiplied by $100/11·2 = 8·95$ to make it correct. Therefore the true current is $1 \times 8·95 = 8·95$ A in this part of the circuit. Figure 8.11(a).

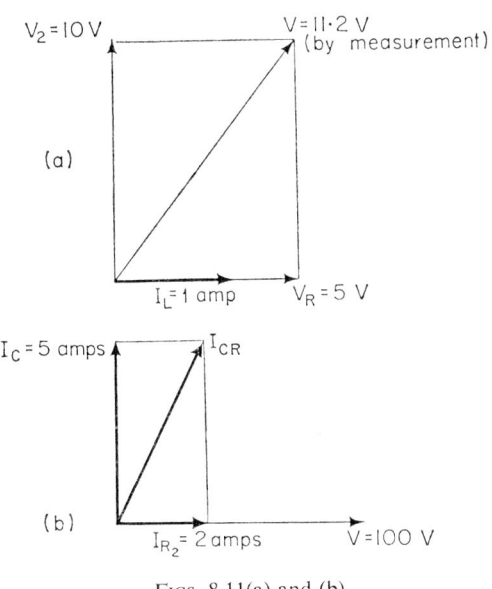

FIGS. 8.11(a) and (b)

(3) Draw a voltage vector of length 100 V.

(4) Draw the vector for the current I_{R_2} in the 50 Ω resistor, value $100/50 = 2$ A, in phase with the voltage vector. Figure 8.11(b).

(5) Draw the vector for the current I_C in the capacitor value

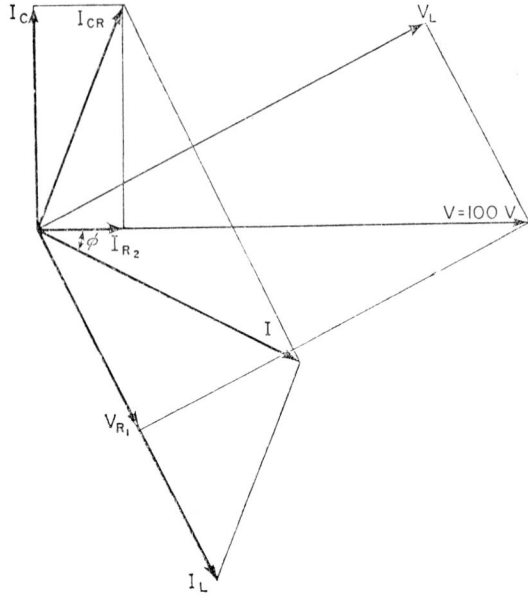

FIG. 8.11(c)

100/20 = 5 A, 90° in advance of the voltage vector. Figure 8.11(b).

(6) Combine currents I_C and I_R by completing a parallelogram giving the resultant I_{CR}. Figure 8.11(b).

(7) Superimpose the diagram obtained in (2) enlarged so that the resultant voltage from that diagram exactly fits the voltage vector drawn in (3). Current I_L should now measure 8·95 A. Figure 8.11(c).

(8) Combine currents I_{CR} and I_L by completing a parallelogram giving the resultant I. Figure 8.11(c).

The vector diagram is now complete and any information which may be required about the voltages and currents in the various branches may be read off. For instance the current taken from the supply is $I = 6 \cdot 7$ A, the phase angle relative to the applied p.d. is $\phi = 32°$, the current in the capacitor is $I_C = 5$ A, in the 50 Ω resistor is $I_{R_2} = 2$ A and in the inductor is $I_L = 8 \cdot 9$ A. The p.d. across the inductor is $V_L = 89$ V and that across the 5 Ω resistor is 45 V. Thus the vector diagram gives a very complete picture of what is happening in the circuit.

It should be clear that there is no greater inherent difficulty in solving a circuit of many branches compared with that containing few, the main difference being the large number of vectors all starting from a single point which the many branched circuit necessitates. However, by the choice of a reasonably large scale a very considerable number of vectors can be drawn and the construction lines for completing parallelograms can be put in faintly and thus not confused with vectors. It is also quite possible to draw current vectors and voltage vectors in contrasting colours or in ink and pencil thus improving the readability of the diagram; it is in any case usual to distinguish between the arrow heads for current, which may be →, and voltage, which may be →.

We will next consider the use of vectors in the solution of some special circuits which are of common occurrence and where the behaviour of the circuit under conditions of varying reactance or varying resistance may be determined. The vector method is particularly suited to demonstrating the way in which certain circuits work and indicating the best form for them.

As a preliminary it should be noted that any part of a circuit containing a resistance in series with a reactance will result for that part of the circuit in a voltage vector diagram, consisting of a right-angled triangle the hypotenuse of which is the voltage across the combination; similarly if a resistance is in parallel with a reactance the current vectors form a right-angled triangle. This means that a semi-circle can be drawn with the hypotenuse as a diameter as shown in Fig. 8.12 and the intersection point P of the other two vectors must always lie on this semi-circle. Now variation of frequency has a proportional effect on an inductive reactance and an inversely proportional effect on a capacitive reactance as is clear from the equations $X_L = \omega L$ Ω and $X_C = 1/\omega C$ Ω. Thus the effect of change in frequency is to make the vector intersection point travel round the semi-circle reaching a limit at one end of the diameter for zero frequency, i.e. a steady voltage or current, and at the other end for an infinitely high frequency or in practice a frequency sufficiently high for R to be negligible compared with X_L or X_C.

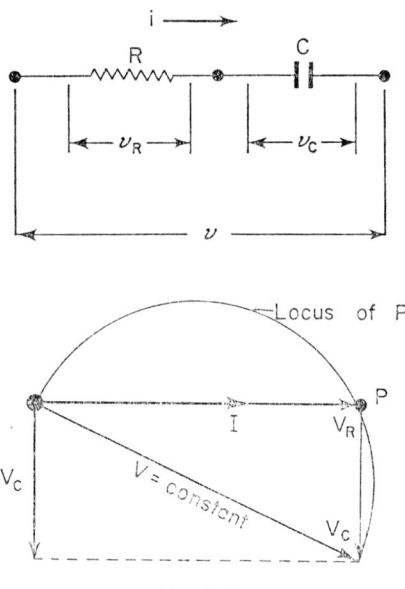

FIG. 8.12

Figure 8.13 shows a circuit similar to some considered in the last chapter which as we have seen can, by adjustment of the value of any one element or by adjustment of the frequency, be put into a resonant state.

The vector diagram for this circuit is shown in Fig. 8.14. In (a) the vector diagram is drawn in the normal manner but in (b) the vector representing the current in the capacitor is reversed thus making the vector sum of I_C and I_L more convenient to draw since all that is necessary is to join the tips of these two vectors. Now the scales for current and voltage are independent, thus a scale can be chosen so that the length of the I_L vector is the same as the V_R vector for all values of I_L. This will be true for all values if it is true for any one value because $I_L R = V_R$ and R

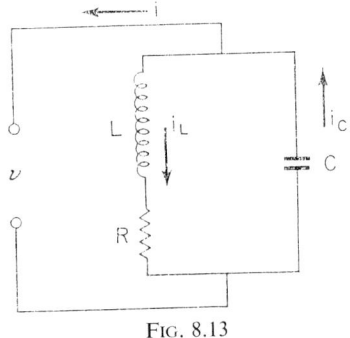

FIG. 8.13

does not vary with frequency. We thus finally get the vector diagram Fig. 8.14(c) representing the circuit.

Now let the value of the capacitor be changed which will result in its reactance being changed and hence the value of the current through it will alter. This has the effect on the diagram of altering the length of I_C but leaving the remainder of the diagram unchanged. It will be seen that the closest approach of the tip of this vector to the top of I_L is the condition shown in Fig. 8.15 where a perpendicular has been dropped from the tip of I_L onto the locus of the tip of I_C. This then is one of the states of resonance and the resultant current can be read from the diagram. The value of I_C is also determined and, since $X_C = V/I_C$, this can now be calculated finally giving $C_{resonance} = 1/\omega X_C$ from which the value of C necessary to produce resonance can be calculated at once. It is also clear that at resonance the current from the source is in phase with the voltage of the source and hence the net reactance of the circuit is zero.

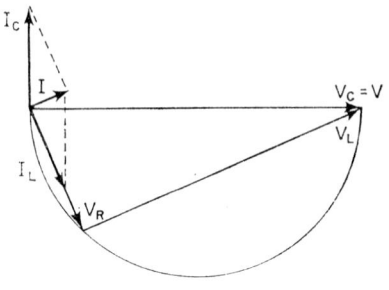

FIG. 8.14(a)

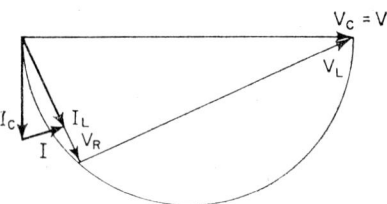

FIG. 8.14(b)

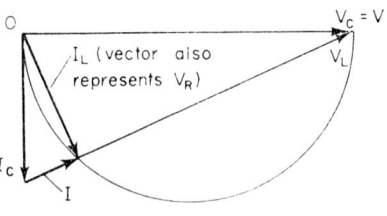

FIG. 8.14(c)

The following relationship can be derived directly from Fig. 8.15 bearing in mind that the current and voltage triangles are similar

$$\frac{I}{I_C} = \frac{V_R}{V_L}$$

$$\therefore \quad I = \frac{V_R I_C}{V_L} = \frac{I_R R \times V \omega C}{\omega L I_R}$$

$$\therefore \quad \text{Impedance} = \frac{V}{I} = \frac{V \times \omega L I_R}{I_R R \times V \omega C} = \frac{L}{RC}$$

L/RC is sometimes known as the dynamic impedance.

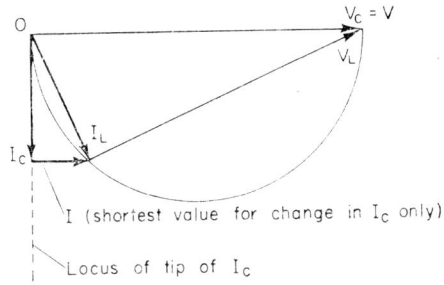

FIG. 8.15

Also if Q is defined as the ratio of reactance at resonance to resistance for the inductor and resistor

$$Q = \frac{\omega L}{R}$$

$$= \frac{\omega L I_R}{R I_R} = \frac{V_L}{V_R} = \frac{I_C}{I}$$

$$\therefore \quad I_C = QI$$

Thus the current in the capacitor is Q times the current taken from the supply.

If Q is large

$$I \ll I_C \quad \text{and} \quad R \ll \omega L$$

$$\therefore \quad I_C \simeq I_L$$

and

$$I_C = \frac{V}{1/\omega C} \quad \text{also} \quad I_L \simeq \frac{V}{\omega L}$$

$$\therefore \quad \frac{V}{1/\omega C} = \frac{V}{\omega L}$$

$$\therefore \quad \frac{1}{\omega C} = \omega L \quad \text{and} \quad \omega = \frac{1}{\sqrt{LC}}$$

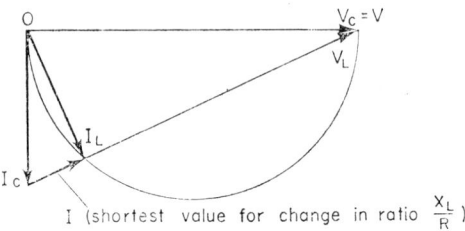

I (shortest value for change in ratio $\frac{X_L}{R}$)

FIG. 8.16

Next suppose that the ratio X_L/R is varied by altering L or R. This will cause the intersection point of V_R and V_L to move round the circle as shown in Fig. 8.16 and it will be seen that the minimum current is obtained when this intersection point lies on the radius to the circle which when produced will pass through the tip of I_C as in Fig. 8.16. This is another condition of resonance where I is a minimum but is not now in phase with V.

Another useful circuit is that shown in Fig. 8.17. This circuit is such that the potential difference between D and B is always

$1/\sqrt{2}$ times that between A and C but simultaneous alteration of the values of the inductor L and resistor r cause the phase of V_{DB} to vary relative to V_{AC} over a wide range. The circuit is thus known as a phase shifting circuit and the component values must satisfy certain conditions which are that

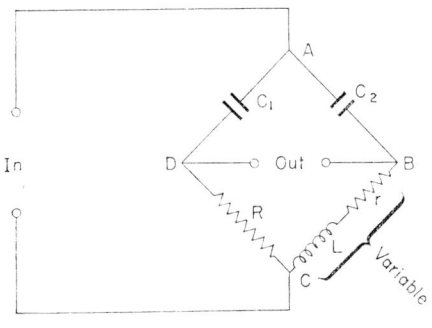

FIG. 8.17

$R = X_{C1}$, that the simultaneous variation of L and r is such that $X_L = r$ for all settings and that C_1 and C_2 are very good capacitors having negligible resistance. If these conditions are met the vector diagram for the circuit may be developed as shown in Fig. 8.18. First draw vector V representing the p.d. of the source. This must be equal to the vector sum of V_R and V_{C1} as shown in Fig. 8.18(a), where angle ϕ must be a right angle. Next draw a vector diagram for V_{CB} which is the vector sum of V_L and V_r. Since X_L must always equal r according to our special conditions, the angle θ, Fig. 8.18(b), must always be 45°. Now vectorially add V_{Lr} to V_{C2} remembering that the current flowing through L_r must also flow through C_2 thus making the direction of vector V_{C2} directly opposite to that of V_L. This gives the result shown in Fig. 8.18(c) and the resultant must be equal to V. Now combine the two vector diagrams making V a common vector giving Fig. 8.18(d). Since angle ϕ is always 90° and angle ϕ is always 45° CBA will lie on a circle of centre D and radius $DC = DA = DB$. But DB is the vector representing

140

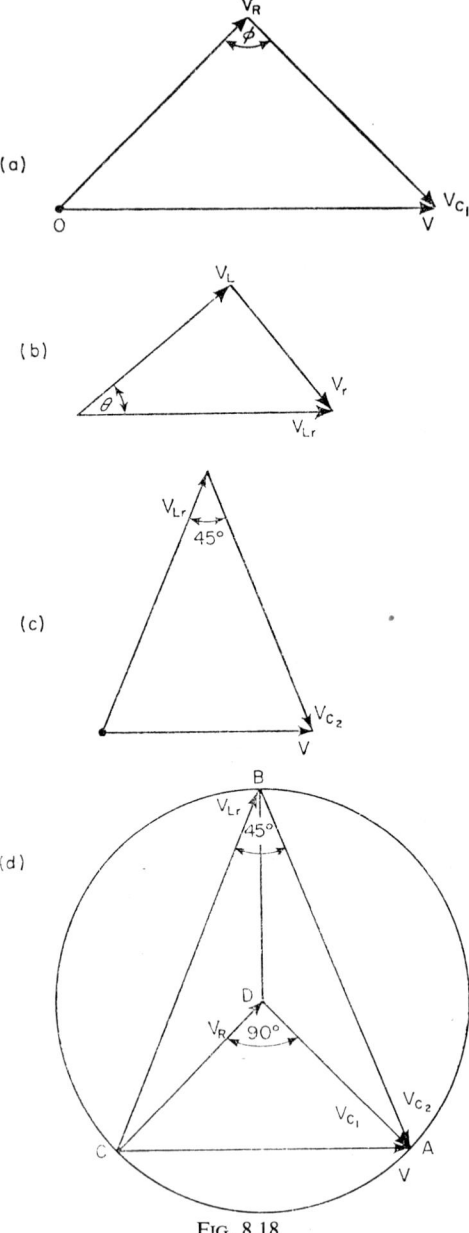

(a)

(b)

(c)

(d)

Fig. 8.18

the potential between D and B and hence this potential remains constant in magnitude but variable in phase relative to potential applied across AC.

Another phase-shifting circuit used for detecting change of frequency is simpler in operation and is shown in Fig. 8.19. Here a change in frequency causes the phase of the p.d. between B and D to alter without change in magnitude. The vector

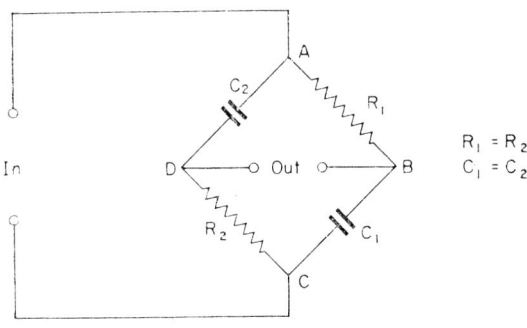

FIG. 8.19

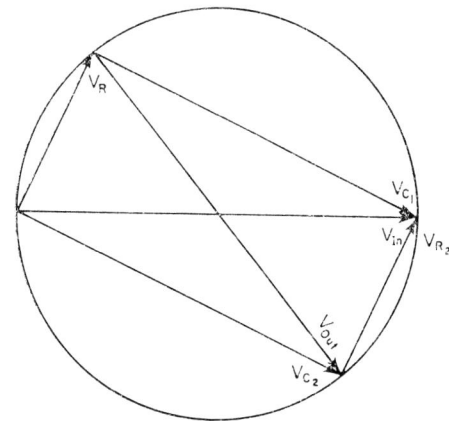

FIG. 8.20

diagram for this circuit is shown in Fig. 8.20 and is less complex than that for the previous circuit. It is clear from the diagram that the vector representing the p.d. *DB* will always be a diameter of the circle and hence will be constant but its phase relative to *AC* will vary depending on the ratio of *AB* to *BC* (or *CD* to *DA*). This ratio is given by the ratio of *R* to X_C and, since $X_C = 1/\omega C$ and *R* is constant, will clearly vary with changes in $f = \omega/2\pi$.

Typical Examination Questions

(The following questions should be solved graphically).

1. A p.d. of 250 V a.c. is maintained across a circuit consisting of a 5 Ω resistance in series with a coil which has a resistance of 8 Ω and a reactance of 6 Ω; a 10 Ω resistance is connected as a shunt across the terminals of the coil. In the steady state, what is the p.d. across the coil?

Ans: 130 V.

2. An a.c. of 1 kA is transmitted by two circuits in parallel; one of these has a resistance of 2 Ω and negligible reactance, the other has a resistance of 2 Ω and a reactance of 2 Ω. Find the current in each circuit.

Ans: 633, 447 A.

3. Figure 1 shows in ohms the resistances and reactances of the branches of a network. If the terminals *A* and *D* are connected to mains of 200 V p.d., find the p.d. across *BD*, the current in *BCD* and their phases relative to the p.d. of the mains.

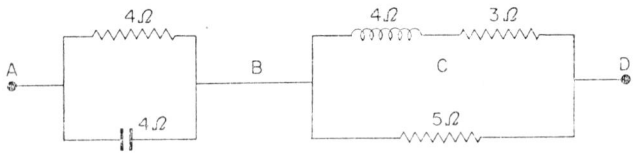

Ans: 122·5 V leading by tan⁻¹ 8/11,
          24·5 A lagging by tan⁻¹ 4/13.

Fig. 1

4. An alternating p.d. of constant rms value *V* and variable frequency *f* is maintained across a load of two parallel branches. One branch is a coil of inductance *L* henries and resistance *R* ohms; the second branch consists

of a non-inductive resistance also of R ohms in series with a condenser of capacitance C farads such that $C = L/R^2$. Show that whatever the frequency f may be, the currents in the two branches are in quadrature with one another, and that the total current is always in phase with V and always has the value V/R.

5. A condenser of capacitance $C\,\mu F$, in parallel with a non-inductive resistance of R ohms, is connected to an alternator giving a constant p.d. of 150 V at various frequencies. At a frequency of 50 c/s the steady current in the joint circuit is 0·08 A; at a frequency of 90 c/s it is 0·10 A. Find C and R.

Ans: 0·85; 2160.

6. (a) A coil of 200 μH inductance has a resistance of 6·28 ohms when measured at a frequency of 1 Mc/s. What is the Q-factor of the coil at this frequency?

Ans: 200.

7. The coil of Q.6 is connected in parallel with a perfect condenser C to form a parallel resonant circuit which is tuned to be purely resistive at 1 Mc/s. What is then the value of this resistance?

Ans: 0·251 MΩ.

CHAPTER 9

Symbolic Method of Analysis

THE solution of a.c. circuits supplied from sinusoidal sources is much facilitated by the use of vector methods and a graphical method of using vectors constitutes a very neat and concise solution to many circuit problems. However, it is not always possible to construct the vector diagram appropriate to a specific problem without using trial and error constructions, since the formulation of the problem may contain too many unknowns. In these problems and in others for which graphical methods are in any case not very suitable the j operator method should be used.

The aim of the j operator method is to express the vector in mathematical form so that the ordinary rules of algebra may be applied to it, and hence the results of addition, subtraction, multiplication and division of vectors can be obtained without having to draw any of them.

Figure 9.1 shows a vector of length $OA = r$ at an angle θ to the x axis of the coordinate axes. It is clear that this vector is defined equally well by $x = a$ and $y = b$ which are the coordinates of the point A and the vector OA is also the vector sum $OB + BA$ where the length of $OB = a$ and the length of $BA = b$. Now the vector OA could be in any quadrant of the coordinate system so it is clear that any vector originating at O, whatever its direction, could be specified as the sum of two components, one measured along the x–x axis and the other along the y–y axis.

The essence of the j operator method is that it combines in one algebraic expression the two mutually perpendicular components of any vector originating at O and we achieve this by stating that all quantities measured in the y direction are to be

144

multiplied by j, all other quantities measured in the x direction remaining unaltered. Thus the vector OA will now become $(a + jb)$.

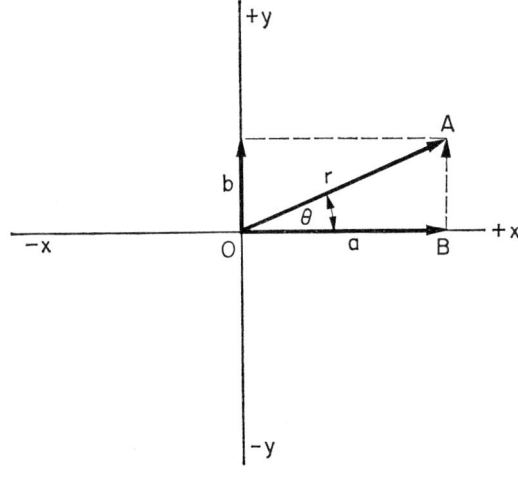

FIG. 9.1

Figure 9.2 shows a vector in each of the four quandrants and their designation in the j notation are $OA = (+a + jb)$, $OB = (-c + jd)$, $OC = (-e - jf)$ and $OD = (+g - jh)$.

It will be clear that if the multiplication of a vector by j means that its direction is that of the y axis then any vector originally unmultiplied by j and lying along the x axis could be turned so as to lie along the y axis by multiplying it by j. Hence it can be said that multiplying a vector by j turns it by 90° in a counter-clockwise direction and this is true of any vector whatever since all vectors can be split up into components along the x and y axes.

Further multiplication by j continues to rotate the vector 90° each time, so that starting with a vector of length a and lying along the $+x$ direction we have ja as a vector lying along the

$+y$ direction, j^2a a vector lying along the $-x$ direction and j^3a a vector lying along the $-y$ direction. Now the vector of length a lying along the $-x$ direction is $(-1) \times a$, thus multiplication by j^2 is equivalent to multiplication by (-1). Hence j is frequently said to be equal to $\sqrt{(-1)}$ and to be an imaginary quantity.

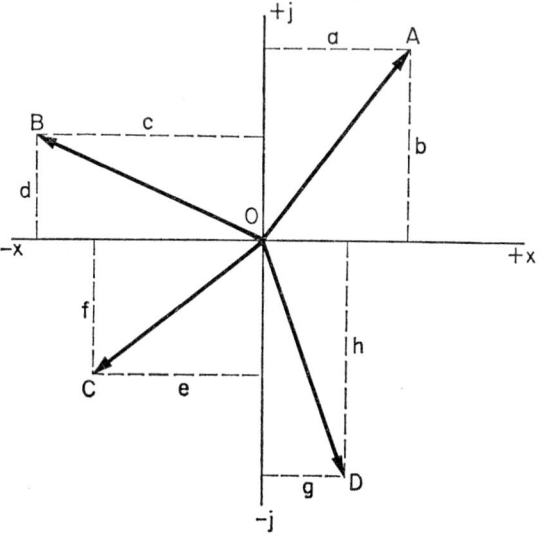

FIG. 9.2

Similarly, a vector of length a lying along the $-y$ axis is $(-1) \times ja = -ja$ and thus $j^3 = -j$. Finally, it should be noted that $1/j = j/j^2 = j/(-1) = -j$.

To distinguish vector quantities from scalar quantities in printed matter it is usual to print the symbol for the vector in bold type frequently as a capital letter. Thus the vector in Fig. 9.1 would be printed $\mathbf{R} = (a + jb)$ thus distinguishing it from $r = (a^2 + b^2)^{\frac{1}{2}}$ which follows from the geometry of the figure. $(a + jb)$ is known as a complex quantity having a real part a and an imaginary part jb, and in any equation involving complex quantities the real parts on each side of the equation must be

equal separately from the imaginary parts, which must also be
equal.

Since the great advantage of the j method of solving circuit
problems is the ease with which the expressions for the vectors
can be manipulated algebraically, we must next consider the
various ways in which the vectors can be expressed so that the
most suitable form may be chosen for any particular problem.

The vector $\mathbf{R} = r\underline{/\theta}$ (i.e. length r at any angle θ as shown in
Fig. 9.1) may be written in j notation in the following ways:

(i) $\mathbf{R} = (a + jb)$ where $a = r \cos \theta$ and $b = r \sin \theta$

(ii) $\mathbf{R} = r \cos \theta + jr \sin \theta = r(\cos \theta + j \sin \theta)$

(iii) $\mathbf{R} = r e^{j\theta}$

The last of these expressions is not obvious and it can be
proved as follows:

$$\cos \theta = 1 - \frac{\theta^2}{2!} + \frac{\theta^4}{4!} - \frac{\theta^6}{6!} + \dots$$

$$\sin \theta = \theta - \frac{\theta^3}{3!} + \frac{\theta^5}{5!} - \frac{\theta^7}{7!} + \dots$$

$$e^{\theta} = 1 + \theta + \frac{\theta^2}{2!} + \frac{\theta^3}{3!} + \dots$$

$$j^2 = -1, j^3 = -j, j^4 = 1, j^5 = j, j^6 = -1, \text{ etc.}$$

$$\therefore e^{j\theta} = 1 + j\theta - \frac{\theta^2}{2!} - \frac{j\theta^3}{3!} + \frac{\theta^4}{4!} + \frac{j\theta^5}{5!} - \frac{\theta^6}{6!} - \frac{j\theta^7}{7!} + \dots$$

$$= \left(1 - \frac{\theta^2}{2!} + \frac{\theta^4}{4!} - \frac{\theta^6}{6!} + \dots\right) + j\left(\theta - \frac{\theta^3}{3!} + \frac{\theta^5}{5!} - \frac{\theta^7}{7!} \dots\right)$$

$$\therefore \mathbf{R} = r(\cos \theta + j \sin \theta) = re^{j\theta}$$

We must next consider the four basic algebraic processes addition,
subtraction, multiplication and division as applied to vectors in
the j rotation.

Suppose we have two vectors $A = a + jb$ and $B = c + jd$ as shown in Fig. 9.3(a) and that we wish to add them. The graphical solution is shown in Fig. 9.3(b) where it is clear that the resultant vector C is the sum of two components $(a + c)$ parallel to the $+x$ axis and $(b + d)$ parallel to the $+j$ axis. Then the sum of the two vectors is given by $C = (a + c) + j(b + d)$. This is obviously

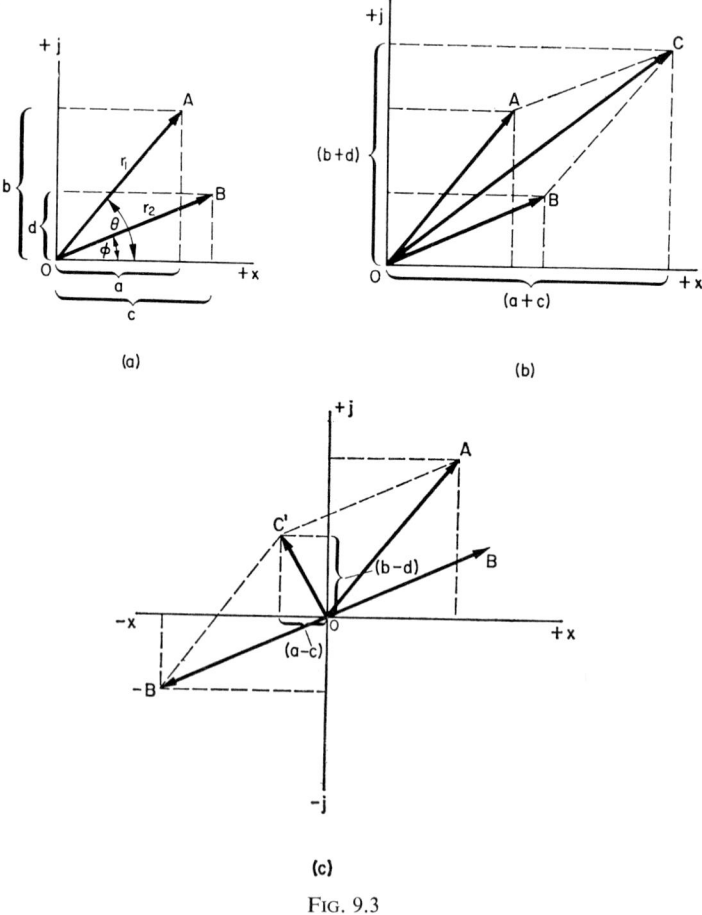

(a)

(b)

(c)

FIG. 9.3

the same as $C = (a + jb) + (c + jd)$ except that the real and imaginary parts have been collected together. In a similar manner the same two vectors may be subtracted as shown in Fig. 9.3(c) where vector B is subtracted from vector A by drawing B in reverse direction (i.e. multiplying it by -1) and adding the result giving components for the resultant vector of $(a - c)$ and $(b - d)$. This is clearly a vector $C' = (a - c) + j(b - d)$ and is obviously the same as $C' = (a + jb) - (c + jd)$.

Vectors may also be multiplied together and to understand the effect of this it is best to consider the operation in two parts. Multiplication by a real number simply increases the magnitude of a vector since the real component and the j component are increased by the same factor. Multiplication by j turns a vector through 90° counter-clockwise. Thus the result of multiplying one vector $A = (a + jb)$ by another $B = (c + jd)$ can be thought of as first increasing A by the factor c and then adding to this result a vector of length $d \times A$ at right angles to the original vector A. Algebraically this is as follows:

$$(a + jb) \times (c + jd) = ac + jad + jbc + (j)^2 bd$$
$$= ac + jad + jbc - bd \text{ (since } j^2 = -1)$$
$$= (ac - bd) + j(ad + bc)$$

Vectors may be divided in a similar manner but a little extra manipulation is required so that the answer may be presented in the usual form of a real component and a j one. The working is as follows:

$$\frac{(a + jb)}{(c + jd)} = \frac{(a + jb) \times (c - jd)}{(c + jd) \times (c - jd)}$$

[i.e. multiplying top and bottom of the fraction by $(c - jd)$]

$$= \frac{(a + jb) \times (c - jd)}{c^2 - j^2 d^2}$$

$$= \frac{(a + jb) \times (c - jd)}{c^2 + d^2} \text{ since } -j^2 = +1$$

It will be noticed that there are now no j terms in the denominator and this process is known as rationalization. Its effect is to give us a real number with which to divide the numerator so that we can proceed:

$$= \frac{1}{c^2 + d^2} [ac - jad + jbc - (j^2)bd]$$

$$= \frac{1}{c^2 + d^2} (ac - jad + jbc + bd)$$

$$= \frac{1}{c^2 + d^2} [(ac + bd) + j(-ad + bc)]$$

and this is the resulting vector.

The answer can be obtained more rapidly if the exponential form of the vectors is used. The expressions are thus:

$$\frac{r_1 e^{j\theta}}{r_2 e^{j\phi}} = \frac{r_1}{r_2} e^{j(\theta - \phi)}$$

where r_1 and r_2 are the length of the vectors and θ and ϕ are their phase angles as shown in Fig. 9.3(a).

The length of a vector such as r_1 or r_2 is often referred to as its modulus and gives the actual value of the current or voltage. Its angle, such as θ or ϕ, is called its argument.

The examples given below show how the commonest types of expression met with in electrical networks may be solved. We shall evaluate, for vectors $A = (1 + j2)$ and $B = (3 + j4)$, the following expressions; $A + B$, $A - B$, AB, A/B, A^2, B^3, giving resultant, modulus, argument, and polar form in each case.

(1) $A + B$ $= (1 + j2) + (3 + j4)$

               $= (1 + 3) + j(2 + 4) = 4 + j6$

∴ resultant $= (4 + j6)$

     modulus $= \sqrt{(4^2 + 6^2)} = 7 \cdot 2$

     argument $= \tan^{-1} \dfrac{6}{4} = 56°$

     polar form $= 7 \cdot 2 \underline{|56°}$

(2) $A - B$ $= (1 + j2) - (3 + j4)$

$\qquad\qquad = (1 - 3) + j(2 - 4) = -2 - j2$

$\therefore$ resultant $= -(2 + j2)$

modulus $= \sqrt{(2^2 + 2^2)} = 2\cdot83$

argument $= \tan^{-1} \dfrac{-2}{-2} = 225°$ (not $45°$ because both real and j components are negative)

polar form $= 2\cdot83 \underline{|225°}$

(3) AB $= (1 + j2)(3 + j4)$

$\qquad\qquad = 3 + j4 + j6 + j^2 8$

$\qquad\qquad = 3 + j4 + j6 - 8$

$\qquad\qquad = (3 - 8) + j(4 + 6) = -5 + j10$

$\therefore$ resultant $= -(5 - j10)$

modulus $= \sqrt{[5^2 + (-10)^2]} = 11\cdot19$

argument $= \tan^{-1} \dfrac{+10}{-5} = 116°$

polar form $= 11\cdot19 \underline{|116°}$

(4) A/B $= \dfrac{(1 + j2)}{(3 + j4)}$

$\qquad\qquad = \dfrac{(1 + j2)(3 - j4)}{(3 + j4)(3 - j4)}$

$\qquad\qquad = \dfrac{3 - j4 + j6 - j^2 8}{3^2 - j^2 4^2} = \dfrac{(3 + 8) + j(-4 + 6)}{9 + 16}$

$\qquad\qquad = \dfrac{11 + j2}{25} = 0\cdot44 + j0\cdot08$

$\therefore$ resultant $= (0\cdot44 + j0\cdot08)$

modulus $= \sqrt{(0\cdot44^2 + 0\cdot08^2)} = 0\cdot45$

argument $= \tan^{-1} \dfrac{0\cdot08}{0\cdot44} = 10°$

polar form $= 0.45\underline{|10°}$

(5) A^2 $= (1+j2)^2$

$= (1+j2)(1+j2) = 1+j2+j2+(j2)^2$

$= 1+j4-4 = -3+j4$

∴ resultant $= (-3+j4)$

modulus $= \sqrt{[(-3)^2+4^2]} = 5$

argument $= \tan^{-1}\dfrac{4}{-3} = 126°$

polar form $= 5\underline{|126°}$

Alternatively $A = 1 + j2 = re^{j\theta}$

where $r = \sqrt{(1^2 + 2^2)} = 2.24$

$\theta = \tan^{-1}\dfrac{2}{1} = 63°$

$\therefore A^2 = r^2 e^{j2\theta} = r_1 e^{j\theta}$

where $r_1 = r^2$ and $\theta_1 = 2\theta$

$\therefore r_1 = 2.24^2 = 5 = \text{modulus}$

$\theta_1 = 2 \times 63° = 126° = \text{argument}$

which is the same result as by the other method.

(6) B^3 $= (re^{j\theta})^3$

(where $r = \sqrt{(3^2+4^2)} = 5$, $\theta = \tan^{-1}\dfrac{4}{3} = 53°$)

$= r^3 e^{j3\theta} = 125 e^{j\,159°}$

∴ modulus $= 125$

argument $= 159°$

polar form $= 125\underline{|159°}$

This solution is very much quicker than multiplying out $(3 + j4)^3$. It will be noticed that raising a vector to any power n results in its length being increased by the power n and its phase angle being advanced n times.

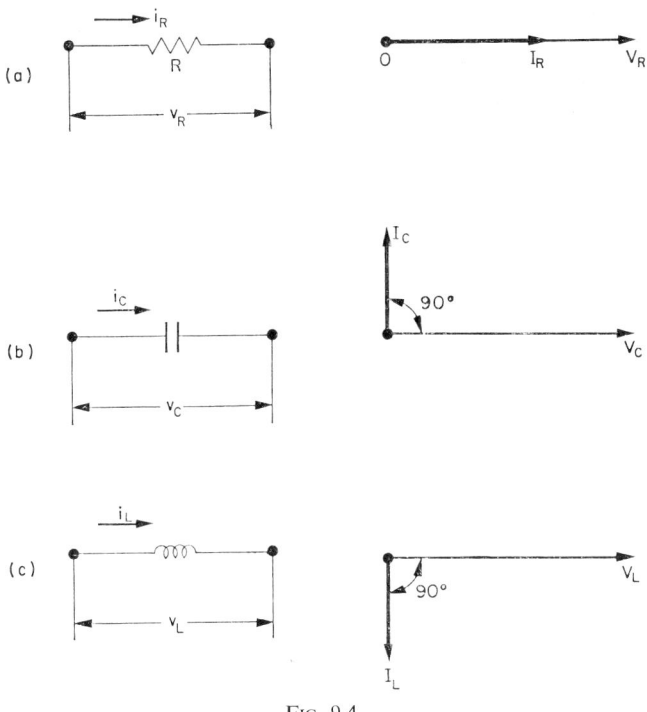

FIG. 9.4

We are now in a position to apply the j method to the solution of problems concerning electrical networks.

Suppose that an ideal resistor carries a sinusoidal alternating current i_R as shown in Fig. 9.4(a). Then there will be a voltage v_R across the terminals of the resistor and the vector diagram for the circuit will be as shown, the vector V_R being in phase with I_R and being of magnitude RI_R.

Similarly, if ideal reactive components are considered the appropriate vector diagrams are as shown in Fig. 9.4(b) and (c) V_c being equal to IX_c but lagging by 90° and V_L being equal to IX_L but leading by 90°.

Now the analytical method must give the same result as the graphical one and we have shown that multiplication by j turns a vector counter-clockwise by 90° and multiplication by $-$j turns it 90° clockwise. Thus $V_c = 0 - jI_cX_c$ and $V_L = 0 + jI_LX_L$ and for the resistor $V_R = I_RR + j0$ since in this case the voltage and current are in phase.

The expressions $0 - jI_cX_c$ and $0 + jI_LX_L$ could be written $I_c(-jX_c)$ and $I_L(+jX_L)$ in which case we have associated the j with the reactance X_c or X_L respectively; the capacitor and the reactor can then be said to have vector reactances of $-jX_c$ and $+jX_L$ respectively. Since $X_c = 1/\omega C = 1/2\pi fC$ and $X_L = \omega L = 2\pi fL$ the vector reactances are numerically $-j/\omega C$ and $+j\omega L$.

When a circuit contains more than one component the application of the results given above produces a j-notation expression known as the *vector impedance* of the circuit. Thus in Fig. 9.5

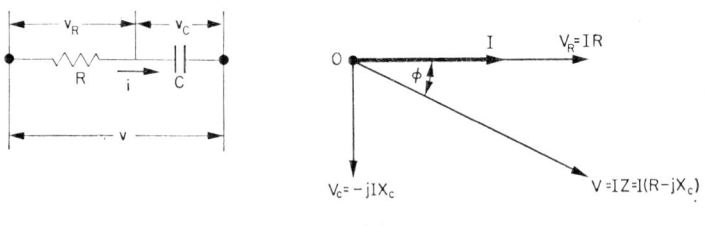

FIG. 9.5

we have a resistor in series with a capacitor giving the vector diagram shown. The resistance of the resistor is $R\ \Omega$, the reactance of the capacitor will be $-jX_c\ \Omega$ in the j notation and the vector impedance of the circuit will be $Z = R + (-jX_c) = (R - jX_c)\ \Omega$. If a current represented by vector I flows through this circuit then the resulting potential across the resistor will be $V_R = IR$, the potential across the capacitor will be $V_c = I(-jX_c) = -jIX_c$ and the potential across the whole circuit will be $V = IR + (-jIX_c) = I(R - jX_c) = IZ$, where $Z = R - jX_c$ and is the impedance of the circuit expressed in vector form.

Application of this method of solution to circuits containing more components is simply a matter of more complex equations but with no new principle involved as the following example will show.

Figure 9.6 gives a circuit having three branches in parallel and connected to a sinusoidal a.c. source at 100 V p.d. We wish to find the current drawn from this source.

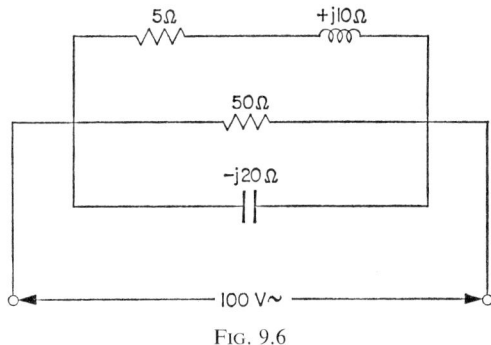

FIG. 9.6

Impedance of top branch $\quad = Z_1 = 5 + j10 \ \Omega$

Impedance of middle branch $= Z_2 = 50 \ \Omega$

Impedance of bottom branch $= Z_3 = -j20 \ \Omega$

If the impedance of the whole circuit is Z then

$$\frac{1}{Z} = \frac{1}{Z_1} + \frac{1}{Z_2} + \frac{1}{Z_3}$$

$$\frac{1}{Z_1} = \frac{1}{5 + j10} = \frac{1 \times (5 - j10)}{(5 + j10)(5 - j10)} = \frac{5 - j10}{5^2 + 10^2}$$

$$= (0 \cdot 04 - j0 \cdot 08) \ \Omega$$

$$\frac{1}{Z_2} = \frac{1}{50} = 0\cdot02$$

$$\frac{1}{Z_3} = \frac{1}{-j20} = \frac{j}{-j^220} = +j0\cdot05$$

$$\therefore \quad \frac{1}{Z} = (0\cdot04 - j0\cdot08) + 0\cdot02 + j0\cdot05$$

$$= 0\cdot06 - j0\cdot03$$

but $\quad I = \dfrac{V}{Z}$

$$\therefore \quad I = 100(0\cdot06 - j0\cdot03) = 6 - j3 \text{ A}$$

$\therefore$ modulus of $I = \sqrt{(6^2 + 3^2)} = \sqrt{45} = 6\cdot7$ A

and this is the value required. The modulus of a vectorial expression is often indicated by enclosing the symbol for the expression between two vertical lines | | and hence modulus of I would be written $|I|$.

A rather more complicated circuit is given in Fig. 9.7, where three branches in parallel are connected to a further branch in series. The solution of the circuit is as follows:

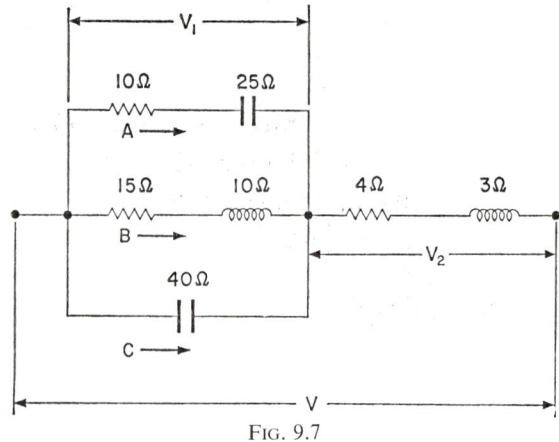

FIG. 9.7

Let the currents in the parallel part of the circuit be represented by vectors A, B and C as shown. Then the current in the series part will be the vector sum $A + B + C$. Let V be 100 volts.

$$V_1 = A(10 - j25) = B(15 + j10) = C(-j40)$$

$$V_2 = (A + B + C)(4 + j3)$$

$$V = V_1 + V_2 = 100$$

$$\therefore 100 = (A + B + C)(4 + j3) - jC40$$

$$= (A + B)(4 + j3) + C(4 - j37)$$

$$A = \frac{-jC40}{10 - j25} = \frac{-jC40(10 + j25)}{100 + 625} = C\left(\frac{1000 - j400}{725}\right)$$

$$= C\left(\frac{40 - j16}{29}\right)$$

$$B = \frac{-jC40}{15 + j10} = \frac{-jC40(15 - j10)}{225 + 100} = C\left(\frac{6400 - j600}{325}\right)$$

$$= C\left(\frac{16 + j24}{13}\right)$$

$$A + B = C\left(\frac{520 - 464 - j208 - j696}{29 \times 13}\right) = C\left(\frac{56 - j904}{29 \times 13}\right)$$

$$= C(0\cdot149 - j2\cdot4)$$

$$(A + B)(4 + j3) = C(0\cdot594 + 7\cdot2 - j9\cdot6 + j0\cdot445) = C(7\cdot794 - j9\cdot155)$$

$$\therefore C = \frac{100}{(11\cdot79 - j46\cdot15)} = \frac{100(11\cdot79 + j46\cdot15)}{2269} = 0\cdot52 + j2\cdot03$$

$$\therefore |C| = \sqrt{(0\cdot52^2 + 2\cdot03^2)} = \sqrt{4\cdot39} = 2\cdot09 \text{ A}$$

$$A = (0\cdot52 + j2\cdot03)(1\cdot38 - j0\cdot552) = 1\cdot837 + j2\cdot513$$

giving $|A| = 3\cdot11$ A

$$B = \frac{(0\cdot52 + j2\cdot03)(16 + j24)}{-13} = 3\cdot1 - j3\cdot46$$

giving $|B| = 4 \cdot 65$ A

$A + B + C = 5 \cdot 46 + j1 \cdot 08$

giving $|(A + B + C)| = 5 \cdot 56$ A

This completes the solution of the currents in the circuit. If their relative phases are required these can be obtained easily from the relationship $\theta = \tan^{-1}(b/a)$ for a vector of form $(a + jb)$ as explained earlier.

The vector expression for the supply current obtained from the above working could be thought of as consisting of two components, one of $5 \cdot 46$ A in-phase with the applied p.d. and the other of $1 \cdot 08$ A leading it by $90°$. If a current component is out of phase with the associated voltage by $90°$ the mean product of the quantities is zero so that we can say that quadrature components of current give no net power. The power associated with the in-phase component is simply the product of that component and the applied voltage. Thus the power drawn from the supply by the above circuit Fig. 9.6 is simply $100 \times 5 \cdot 46 = 546$ W. The ratio of this power to the product of the value of the total current, i.e. $(A + B + C) = 5 \cdot 56$ A, and the voltage applied is known as the power factor of the circuit and is thus

$$\frac{546}{100 \times 5 \cdot 56} = 0 \cdot 982$$

The effect of mutual inductance may be dealt with very readily by the j method. Figure 9.8 shows two coils, of self inductance

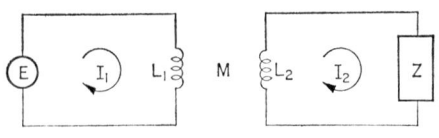

Fig. 9.8

L_1 and L_2 respectively, coupled together to give mutual inductance M. Assuming that these coils are ideal and do not have resistance or capacitance the equations for the two circuits are

For circuit 1: $j\omega L_1 I_1 \pm j\omega M I_2 = E$

For circuit 2: $j\omega L_2 I_2 \pm j\omega M I_1 + Z I_2 = 0$

where it will be seen that we have assumed that the effect of mutual inductance is to insert into one circuit an e.m.f. equal to that which the current in the other circuit would develop across an inductor of value $M\,\Omega$ if it were flowing through it. Hence if there is mutual inductance between two circuits we shall expect to see a $j\omega M I$ term in each of the circuit equations, the current I being that in the "other" circuit in each case. It is helpful to think of ωM as a mutual reactance in ohms.

It is obvious that the phase of $j\omega M I$ relative to the other terms in the equation is very important and the $\pm$ signs have been inserted in the equations above to indicate that there are two ways round of connecting a coil and two ways of showing the directions of I_1 and I_2. It may be agreed that, since E is a sinusoidal source and the currents I_1 and I_2, therefore, are alternating, the direction of the current arrows has no meaning. It is true that if the currents were flowing as shown at the moment then they would be flowing in the opposite direction half a cycle later but the important point to notice is that they would both change and hence the relative direction would remain the same.

The way the coils are wound physically relative to one another is sometimes shown by means of a perspective drawing but it is much commoner for similar ends of the windings to be marked with dots as shown in Fig. 9.9 which indicate that these ends of the coils have the same polarity relative to their respective unmarked ends at any instant. These similar ends are called

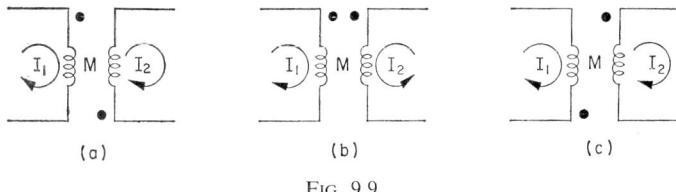

(a)   (b)   (c)

FIG. 9.9

corresponding points and, assuming that M itself is always a positive quantity, $j\omega MI$ will be positive when I_1 and I_2 *both* enter their respective windings at corresponding points and negative in other cases.

This is illustrated in Fig. 9.9, where $j\omega MI$ is positive in every instance, (a), (b) and (c), because I_1 and I_2 *both* enter their respective windings at the dot end in (a) and (b) and *both* enter their windings at the end opposite to the dot in (c).

Typical Examination Questions

1. Obtain the symbolic representation for the following voltages and currents taking $\cos \omega t$ as $1 + 0j$.

(i) $340 \cos \omega t$ V,   (ii) $340 \sin \omega t$ V,

(iii) $340 \cos\left(\omega t + \dfrac{\pi}{6}\right)$ V,   (iv) $340 \cos\left(\omega t - \dfrac{\pi}{6}\right)$ V,

(v) $25 \cos\left(\omega t + \dfrac{2\pi}{3}\right)$ A $+ 25 \cos\left(\omega t - \dfrac{2\pi}{3}\right)$ A,

(vi) $25 \cos \omega t$ A $+ 25 \cos\left(\omega t + \dfrac{2\pi}{3}\right)$ A $+ 25 \cos\left(\omega t + \dfrac{4\pi}{3}\right)$ A.

Ans: (i) $340 + 0j$ V; (ii) $0 - 340j$ V; (iii) $294 + 170j$ V;
(iv) $294 - 170j$ V; (v) $25 + 0j$ A; (vi) $0 + 0j$ A

2. Find the analytical expressions for the waveforms of voltages and currents represented in symbolical form as:

(i) $30 + 40j$ V,   (ii) $40 + 30j$ V,
(iii) $-70 + 70j$ V,   (iv) $(-70 + 70j)$ V $+ (70 + 70j)$ V,
(v) $(10 - 15j)$ A $+ (20 - 35j)$ A $- (15 - 80j)$ A.

Ans: (i) $50 \cos(\omega t + 0.93)$ V; (ii) $50 \cos(\omega t + 0.65)$ V;

(iii) $99 \cos(\omega t + 3\pi/4)$ V; (iv) $140 \cos\left(\omega t + \dfrac{\pi}{2}\right)$ V;

(v) $33.6 \cos(\omega t + 1.11)$ A

3. Two valves pass currents $10 \cos(\omega t + 0.1\pi)$ and $-10 \cos(\omega t - 0.1\pi)$ mA through a common resistance. Find the phase angle between the valve currents and the voltage across the resistance.

Ans: $72°$, one leading and one lagging on the voltage

4. Two alternators in parallel supply a load drawing 400 A. This current lags $30°$ on the alternator voltage. One machine is limited to 200 A maximum output, in phase with the voltage. What current must the second then supply, and at what phase angle to the voltage?

Ans: 248 A, lagging $54°$

5. Two a.c. motors are to draw currents $30 - 20j$ A and $10 - 10j$ A from a supply at $340 + 0j$ V. This is obtained from a transformer via wires of total resistance $0.5\ \Omega$. What voltage must be produced by the transformer; and what is then the percentage increase in voltage at the motor end of the wires if both motors are switched off?

Ans: $360 - 15j$ V; 20.3 on 340 or 6 per cent

6. Express the impedance between A and B for each circuit in Fig. 1 in terms of R, C, L, ω. (Assuming no current leaving at X or Y.)

Ans: (a) $R + j\omega L$, (b) $R - j/\omega C$, (c) $R - j/\omega C$, (d) $R + j\omega L/(1 - \omega^2 LC)$

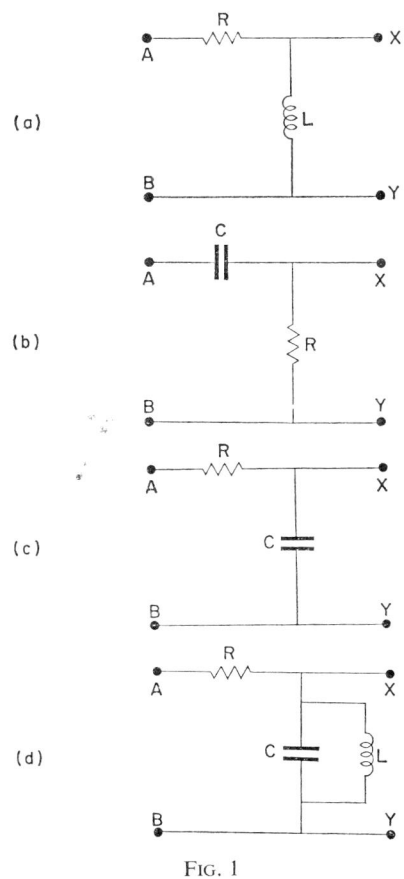

FIG. 1

7. If $R = 1000\ \Omega$, $L = 1/\pi$ H, $C = 1/\pi\ \mu$F, find the ratio of the amplitude of the voltage across XY at 50 c/s to that at 5000 c/s (for the same input voltage across AB); in each circuit in Fig. 1.1. Find also the phase angles between v_{XY} and v_{AB} at each frequency. (Assuming no current leaving X or Y.)

Ans: (a) & (b) 1 : 10, 84° lead, 6° lead,
(c) 10 : 1, 6° lag, 84° lag,
(d) 1 : 1, 84° lead, 84° lag

8. Part of the power supply to a works is at a voltage $340\cos 314t$ V. The current drawn has amplitude 200 A, and it lags 30° on the voltage. By connecting a capacity C across the supply, the amplitude of the total current supplied is reduced. Show that its minimum value is $|i| = 173$; that it will then be in-phase with the supply voltage; and that for this the capacity of the condenser must be 935 μF.

9. Figure 2 shows a circuit used in high frequency amplifiers. Show that

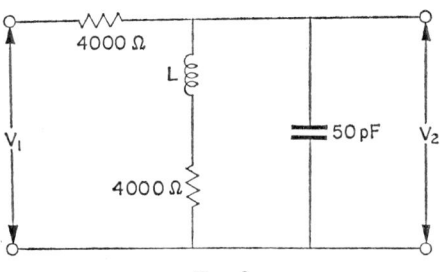

Fig. 2

the amplitude ratio $|V_2|/|V_1|$ falls from 0·5 at low frequencies ($f \ll 1·59$ Mc) to 0·35 at 1·59 Mc, if $L = 0$; but that this drop is reduced to 0·47 if $L = 400$ μH.

CHAPTER 10

Network Theorems

THE first step in solving a network by an analytical method is the formation of an equation or equations to represent the flow of current through the circuit and the disposition of voltages in it. In many instances this may be done directly from the inspection of the circuit diagram but in other instances the circuit may be simplified and the working shortened by the application of network theorems and systematic methods of solution which can be summarized in a set of generally applicable rules.

Let us therefore consider a network consisting of resistors connected in an arbitrary manner to sources of e.m.f. as shown in Fig. 10.1. These resistors are assumed to be of constant value whatever the current flowing through them and whatever the p.d. across them. Hence there will be a relation, $V = IR$, between the voltage and the current where the voltage and current are always proportional in the steady state. This does not preclude the voltage, and hence the current, from being sinusoidal; the important condition is that the value of resistance does not change.

Networks of this type are said to be linear, and they may consist of reactive components provided that the value of reactance is not affected by the value of current or voltage.

Referring to Fig. 10.1 we have assumed that currents I_1, I_2, and I_3 are flowing round the closed circuits $ABDA$, $BCDB$ and $CADC$ respectively. These closed circuits are called meshes and may conveniently be labelled 1, 2 and 3 as shown. A part of one mesh is frequently also part of an adjacent one, and the actual current in the common part is the algebraic sum of the

163

currents assumed in the particular meshes concerned. We then have for mesh 1:

$$R_1(I_1 - I_3) + R_2(I_1 - I_2) + R_4 I_1 = E_1 - E_2 + E_4 \qquad (1)$$

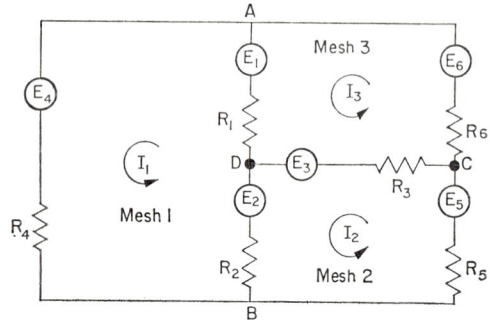

FIG. 10.1

To put the equations more symmetrically

$R_{11} = R_1 + R_2 + R_4 =$ the self-resistance of mesh 1

$R_{12} = - R_2 =$ the resistance shared by meshes 1 and 2

$R_{13} = - R_1 =$ the resistance shared by meshes 1 and 3

and so forth.

Then
$$I_1 R_{11} + I_2 R_{12} + I_3 R_{13} = E_1 - E_2 + E_4$$
$$I_1 R_{21} + I_2 R_{22} + I_3 R_{23} = E_2 - E_3 + E_5 \qquad (2)$$
$$I_1 R_{31} + I_2 R_{32} + I_3 R_{33} = E_3 - E_1 + E_6$$

and
$$R_{13} = R_{31}$$
$$R_{12} = R_{21} \text{ etc.}$$

The set of equations (2) can be solved by determinants to give

$$I_1 = \frac{\begin{vmatrix} (E_1 - E_2 + E_4) & R_{12} & R_{13} \\ (E_2 - E_3 + E_5) & R_{22} & R_{23} \\ (E_3 - E_1 + E_6) & R_{32} & R_{33} \end{vmatrix}}{\begin{vmatrix} R_{11} & R_{12} & R_{13} \\ R_{21} & R_{22} & R_{23} \\ R_{31} & R_{32} & R_{33} \end{vmatrix}} \qquad (3a)$$

and similarly for I_2 and I_3.

When the determinants of the above equations are expanded, the expression for I_1 will be of the form

$$I_1 = a_1 E_1 + a_2 E_2 + a_3 E_3 + a_4 E_4 + a_5 E_5 + a_6 E_6 \qquad (3b)$$

where the a's are constants involving only resistance. In other words, the expression for I_1 contains no terms in which products of the E's occur. The same is true for I_2 and I_3 and we can thus state the following theorem which is known as the Superposition Theorem.

The total current flowing in any part of the circuit is equal to the algebraic sum of the currents which would flow in that part if each of the electromotive forces in turn were acting alone, the other sources being short-circuited.

The determinants which occur in the above equations are completely symmetrical and it is easy to see that, if we had written down the corresponding equations for a network with a larger number of independent loops, we should have obtained for I_1 a similar equation but with determinants of higher order and our proof of the superposition theorem would still have been valid.

Another theorem, the Reciprocity Theorem, states that:

The current I produced in any one branch of a linear network by an e.m.f. E in any other branch, is equal to the current which would be produced in the second branch if E were transferred to the first branch.

By the superposition theorem, the current I produced by E is independent of the currents produced by any other electromotive forces which may be acting in the network so that, in proving the reciprocity theorem, we may neglect these other electromotive forces and currents.

Referring again to Fig. 10.1 the corresponding expression for I_2 is

$$I_2 = \frac{\begin{vmatrix} R_{11} & (E_1 - E_2 + E_4) & R_{13} \\ E_{21} & (E_2 - E_3 + E_5) & R_{23} \\ R_{31} & (E_3 - E_1 + E_6) & R_{33} \end{vmatrix}}{\begin{vmatrix} R_{11} & R_{12} & R_{13} \\ R_{21} & R_{22} & R_{23} \\ R_{31} & R_{32} & R_{33} \end{vmatrix}} \qquad (4)$$

Consider the two branches AB and BC. The current I'_1, in AB resulting from an e.m.f. E in BC is found by putting $E_5 = E$ and $E_1 = E_2 = E_3 = E_4 = E_6 = 0$ in (3a), or

$$I'_1 = \frac{1}{\Delta} \begin{vmatrix} 0 & R_{12} & R_{13} \\ E & R_{22} & R_{23} \\ 0 & R_{32} & R_{33} \end{vmatrix} \tag{5}$$

where Δ is the determinant in the denominators of equations (3a) and (4). Thus

$$I_1 = \frac{E}{\Delta}(R_{13}R_{32} - R_{12}R_{33})$$

Similarly, the current I'_2 in BC resulting from an e.m.f. E in AB is found from (4) to be

$$I'_2 = \frac{1}{\Delta} \begin{vmatrix} R_{11} & E & R_{13} \\ R_{21} & 0 & R_{23} \\ R_{31} & 0 & R_{33} \end{vmatrix} = \frac{E}{\Delta}(R_{23}R_{31} - R_{21}R_{33}) \tag{6}$$

Remembering that

$$R_{13} = R_{31}, \ R_{23} = R_{32} \text{ and } R_{12} = R_{21}$$

we see that

$$I'_1 = I'_2$$

and we have proved the reciprocity theorem for the branches AB and BC. By symmetry it must also be true for any two of the branches AB, BC and CA.

Suppose that, in any network consisting of constant sources of e.m.f. or current and linear resistors, there are two output terminals A and B as shown in Fig. 10.2(a).

Let the open-circuit e.m.f. between the terminals A and B be E, and let the effective resistance of the circuit between them be R. In making a measurement or calculation of R it is assumed that all voltage sources such as E_1 and E_2 are short-circuited, while all current sources such as I_1 are open-circuited.

Thévenin's theorem states that, so far as any external network which may be connected between A and B is concerned, the given network may be replaced by a single e.m.f. E_T in series with a single resistance R_T as shown in Fig. 10.2(b). The proof of this is as follows.

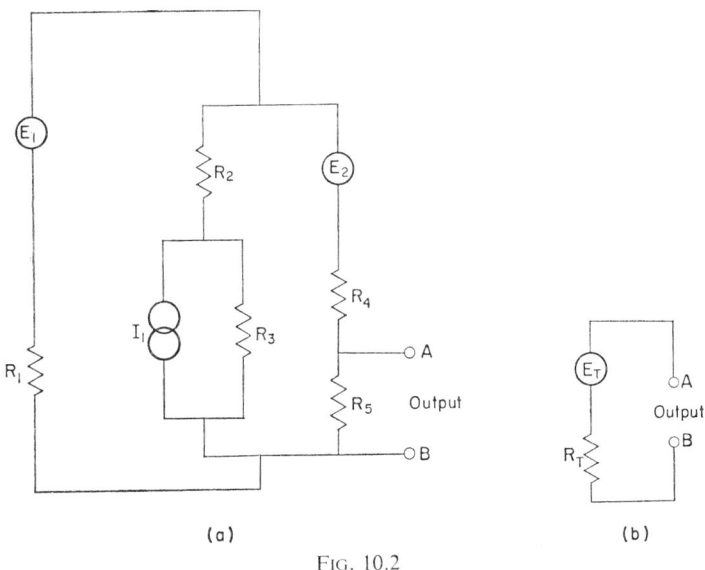

(a)

(b)

FIG. 10.2

In Fig. 10.3 let P represent the given network with terminals A and B, and let Q be the external load network to be connected

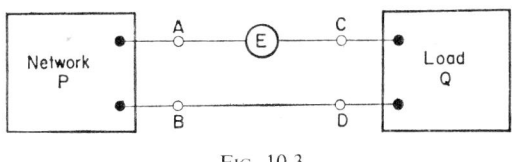

FIG. 10.3

between these terminals. We assume that Q does not contain any sources of e.m.f. or current. This does not restrict the generality of the proof since, by the principle of superposition, the effects of any such sources are independent of currents caused by the source in P. Let the total resistance of Q between C and D be R_1.

Connect between P and Q a generator of e.m.f. E which opposes the open-circuit e.m.f. between A and B. Then the total current flowing between P and Q will be zero. By the superposition theorem, this total current is the sum of (a) the current caused by sources within P, and (b) the current caused by the added generator E. These two currents must be equal and opposite. But the second current is equal to $E/(R + R_1)$ and so, therefore, the first current has the same value in the opposite direction. This proves the theorem, since $E/(R + R_1)$ is the current which would flow between P and Q if P were replaced by an e.m.f. E in series with a resistance R.

A perfect voltage generator is one which maintains a constant voltage between its terminals, irrespective of the current drawn from it. A practical generator can be represented by a constant e.m.f. E in series with a constant resistance R, to allow for loss as shown in Fig. 10.4(a).

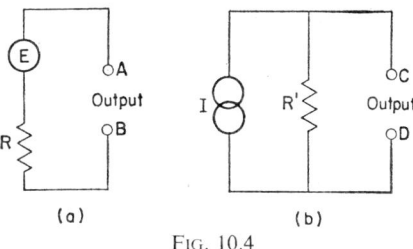

(a)   (b)

FIG. 10.4

It is equally possible to represent a generator as a source of current rather than voltage. A perfect current generator is one which causes a constant current I to flow round an external circuit, irrespective of the resistance of that circuit; thus its internal resistance is infinite. A practical current generator may be represented, as in Fig. 10.4(b), by a perfect current generator I in parallel with a resistance R'.

Generators which can be represented by either of the circuits of Fig. 10.4 are termed linear generators, since they contain only linear elements. By Thévenin's theorem the two circuits will be equivalent, so far as external networks are concerned, if

(i) the open-circuit voltage between A and B is equal to that between C and D;

(ii) the resistance measured at AB is equal to that measured at CD.

Bearing in mind that the resistance of the source E is zero while that of I is infinite, these conditions become

$$R = R'$$

$$E = IR = IR'$$

Thus, whenever it is convenient for purposes of network analysis, a practical linear voltage generator may be replaced by a practical linear current generator or vice versa.

Since a voltage generator E in series with a resistance R is equivalent to a current generator I, of magnitude E/R, in parallel with a resistance R, Thévenin's theorem leads at once to Norton's theorem which states that:

So far as external circuits are concerned, any network with two accessible terminals PQ is equivalent to a constant current generator I in parallel with a resistance R, where I is the current which would flow between P and Q if these terminals were short-circuited and R is the resistance looking back into PQ.

The above theorems have been stated and proved for networks consisting of resistance and direct-current or direct-voltage generators only; however, extensions to a.c. networks can be made without difficulty.

If the direct voltage sources E_1, E_2 ... in Fig. 10.1 be replaced by sinusoidal alternating voltage sources which all have the same frequency, and the resistance R_1, R_2 ... by complex impedances Z_1, Z_2..., the proofs of the superposition theorem and the reciprocity theorem remain valid for steady-state conditions.

If the alternating voltage sources have different frequencies, we cannot use complex impedances because impedance is a function of frequency. However, by writing down the mesh equations in the form of differential equations, we can prove that the superposition theorem is still valid for steady-state conditions.

Since a periodic e.m.f. of any waveform can, by Fourier's theorem, be expressed as the sum of a direct e.m.f. and an infinite series of sinusoidal e.m.f., it follows that both the superposition theorem and the reciprocity theorem are valid for periodic e.m.f. of any waveform, when steady-state conditions prevail.

A sinusoidal alternating voltage generator of e.m.f. E and complex internal impedance Z can be replaced, so far as external circuits are concerned, by a sinusoidal alternating current generator of the same frequency and of magnitude $I = E/Z$, in parallel with an impedance Z.

The proofs of Thévenin's and Norton's theorems remain valid for steady-state conditions if direct current and voltage sources are replaced by sinusoidal alternating sources which all have the same frequency, and if resistances are replaced by complex impedances. Thus the methods of analysis used for d.c. may be adapted for use with a.c. by the use of the j notation method described in Ch. 9.

Numerical examples of these theorems appear, when their use is required, in various worked problems in the remaining chapters of the book; the worked examples given below show how two of the theorems may be applied.

Figure 10.5 shows an ideal current source and an ideal voltage source supplying power to a load R. It is required to find the power delivered by each source and the value of load resistance for which each source would deliver the same amount of power.

By principle of superposition the current in R is the sum of the current which flows when voltage source alone acts, and of the current which flows when the current source alone acts. In each case the other, dead, generator must be left connected.

When I only acts all of its current must flow through the dead voltage generator because it is a short-circuit.

When V alone acts no current can flow through the dead current source for its internal resistance is infinite. A current V/R is driven through R.

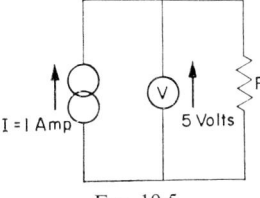

FIG. 10.5

Combine these results. Both sources have a p.d. V across them; a current I flows out of the current source, and a current $(V/R) - I$ out of the voltage source. The powers delivered are therefore $I \times V$ and $[(V/R) - I] \times V$ from the current and the voltage sources respectively. When these are equal, we have:

$$I = \left(\frac{V}{R} - I\right) \quad \therefore \quad R = \frac{V}{2I} = \frac{5}{2 \times 1} = 2 \cdot 5 \ \Omega$$

In the circuit in Fig. 10.6 it is required to find the current in the 1 Ω resistor.

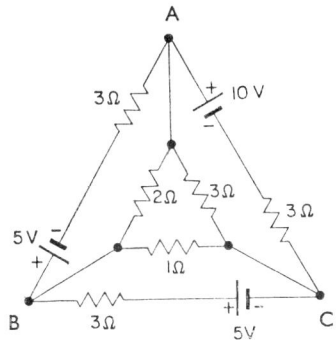

FIG. 10.6

The following four methods could be used:

(a) Thévenin's theorem
(b) Nodal analysis (two unknowns)

(c) Superposition

(d) Mesh analysis (four unknowns)

By method (a) branch AB, shown separately in Fig. 10.7, can be simplified as follows:

In Fig. 10.7 $\qquad I = \dfrac{5}{5} = 1$ A

$\qquad \therefore \ V_{AB} = 2$ V = Thévenin E

R_{AB} (with source of e.m.f. short-circuited)

$$= \frac{3 \times 2}{3 + 2} = 1 \cdot 2 = \text{Thévenin } R$$

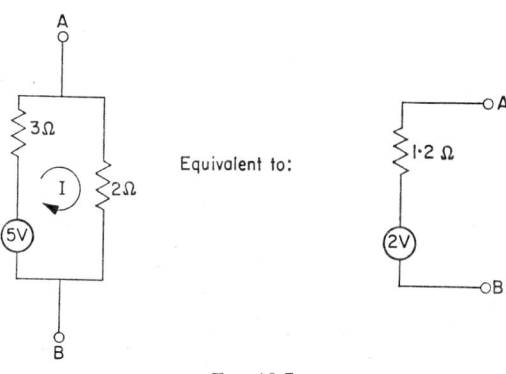

Equivalent to:

FIG. 10.7

Hence the Thévenin equivalent for branch AB is 2 V in series with $1 \cdot 2\ \Omega$.

Similarly we find the equivalent for branch AC is 5 V in series with $1 \cdot 5\ \Omega$ and the equivalent for branch BC is $1 \cdot 25$ V in series with $0 \cdot 75\ \Omega$.

The complete circuit may then be redrawn as in Fig. 10.8 where the circulating current is $5 \cdot 75/3 \cdot 45 = 1 \cdot 66$. The p.d. across the $0 \cdot 75$ resistor is then $0 \cdot 75 \times 1 \cdot 66 = 1 \cdot 25$ V and is additive to the $1 \cdot 25$ e.m.f. in the equivalent for the BC branch. Thus the p.d. across BC is $1 \cdot 25/1 \cdot 25 = 2 \cdot 5$ V and, referring to the original circuit in Fig. 10.6, the current in the $1\ \Omega$ resistor is $2 \cdot 5/1 = 2 \cdot 5$ A.

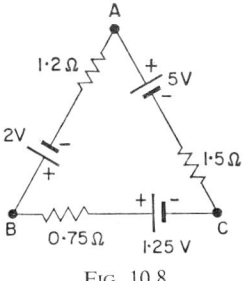

FIG. 10.8

Figure 10.9(a) shows two coils having mutual inductance M, one of which is connected to an alternating source of V volts through a resistor R ohms. It is required to find the Thévenin equivalent of this circuit.

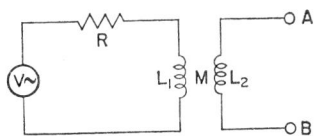

(a)

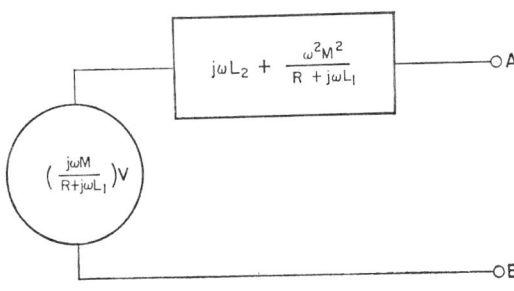

(b)

FIG. 10.9

Open circuit voltage across $AB = V_2 = j\omega M I_1$

$$= j\omega M \frac{V}{R + j\omega L_1}$$

Short circuit current through AB joined together $= I_2$

$\therefore$ Thévenin impedance $= Z_T = \dfrac{V_2}{I_2} = j\omega M \dfrac{I_1}{I_2}$

giving, after some algebra, $Z_T = j\omega L_2 + j\omega M \dfrac{-j\omega M}{R + j\omega L_1}$

$$= j\omega L_2 + \frac{\omega^2 M^2}{R + j\omega L_1}$$

Therefore the Thévenin equivalent is a source of p.d.

$$\left(\frac{j\omega M V}{R + j\omega L_1} \right) \text{V}$$

in series with an impedance of

$$\left(j\omega L_2 + \frac{\omega^2 M^2}{R + j\omega L_1} \right) \Omega$$

as shown in Fig. 10.9(b).

Typical Examination Questions

1. Figure 1 shows two voltage generators driving current round a circuit.

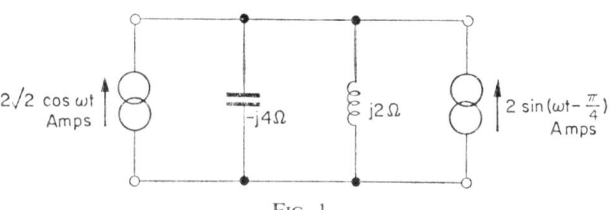

2√2 cos ωt Amps −j4Ω j2Ω 2 sin(ωt − $\frac{\pi}{4}$) Amps

FIG. 1

(a) Express the voltage excitation in complex form.

(b) Set up the circuit equation.

(c) Obtain the modulus and phase angle of I, and then turn the resulting complex quantity back into a cosine.

Ans: (a) $3\,e^{j\omega t}$, $[\sqrt{(3)} - j]\,e^{j\omega t}$,

(b) $I = e^{j\omega t} \times \dfrac{3 - \sqrt{(3)} + j}{1 + j}$,

(c) modulus $1\cdot14$, phase angle $-6\cdot7^\circ$ ($= 38\cdot3^\circ - 45^\circ$)
$1\cdot14\cos(\omega t - 6\cdot7^\circ)$ A

2. Figure 2 shows two current generators feeding a load. Carry out steps similar to (a), (b) and (c) in Question 1, for the node potential. (Hint: first turn the sine into a cosine.)

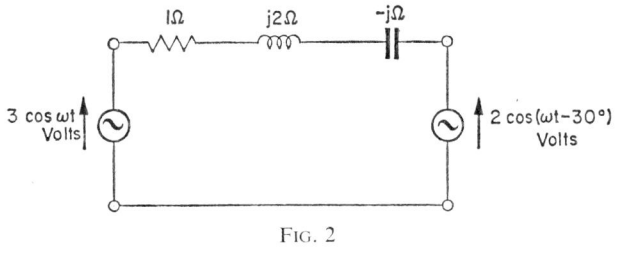

Fig. 2

Ans: (a) $2\sqrt{2}\,e^{j\omega t}$, $-\sqrt{2}(1 + j)\,e^{j\omega t}$,

(b) $V = e^{j\omega t}\,[2\sqrt{2} - \sqrt{2}(1 + j)]/j(\tfrac{1}{4} - \tfrac{1}{2})$,

(c) $8,\ 45^\circ$; $8\cos(\omega t + 45^\circ)$ V

3. Figure 3 shows a circuit with voltage excitations represented by 5 and $(5 + j)$ V acting as shown. The impedances are all in ohms. Find the magnitude and phase of the p.d. across AB. (Hint: by superposition break up the excitation and avoid solving simultaneous equations.)

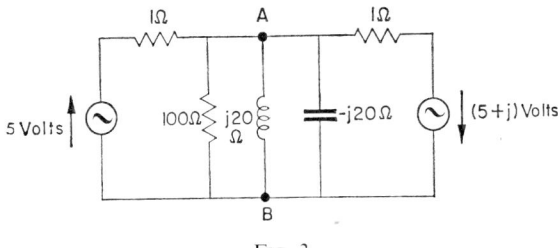

Fig. 3

Ans: Complex potential is $\dfrac{100j}{201}$; $0\cdot497$ V, -90°

4. Represent the excitations in Fig. 4 by complex exponentials. Then set up the mesh equations for I_1 and I_2. Solve for I_2 and turn the complex current back into a physical one.

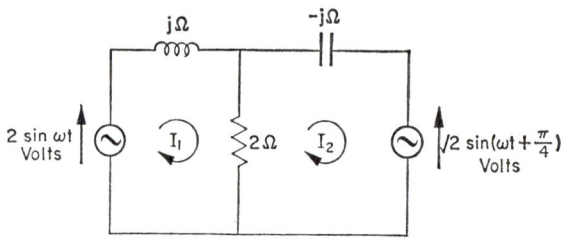

Fig. 4

Ans: $2\,e^{j\omega t}$, $(1+j)\,e^{j\omega t}$

Mesh equations $2\,e^{j\omega t} = (2+j)I_1 - 2I_2$

$-(1+j)\,e^{j\omega t} = 2I_1 + (2-j)I_2$

I_2: $3(1-j)\,e^{j\omega t}$ or $3\sqrt{2}\sin(\omega t - \pi/4)$ A

5. The circuit of Fig. 5 is excited at an angular frequency of 1000 rad/sec. Find the magnitude of the p.d. across the 8000 Ω load, and the possible phase angles with respect to the excitation.

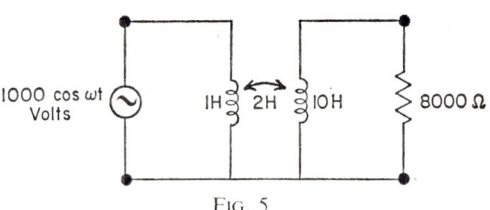

Fig. 5

Ans: $\pm 1600\cos(\omega t + 143°)$ **volts**

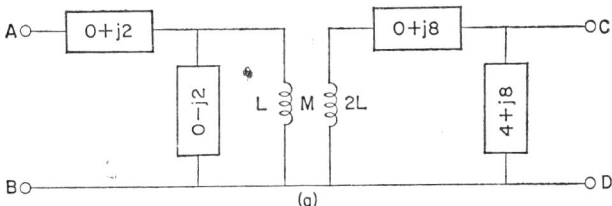

(a)

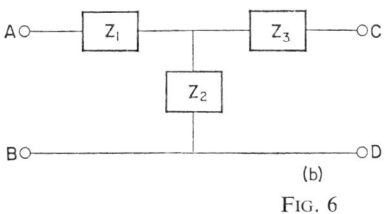

(b)

FIG. 6

6. Find the impedances Z_1, Z_2, Z_3, which make Fig. 6(b) equivalent to Fig. 6(a) given that $M = \sqrt{(2)}L$.

$$Ans: Z_1 = \frac{4}{1+2j} \pm 4j; \; Z_2 = \pm 4j; \; Z_3 = \pm 4j$$

CHAPTER 11

Bridge Networks

AN ELECTRICAL bridge network is an assembly of electrical components connected in such a way as to facilitate the measurement of some desired quantity or the comparison of two quantities. It is often desirable that a small change in the parameter being measured shall produce a large change in the detector or measuring device and also that the stability and accuracy of the bridge shall be good. All bridge networks can be treated for analysis purposes as being assemblages of ideal components. Their circuit diagrams do not differ from any other electrical network and hence normal methods of circuit analysis are used in their solution. The quantity that is usually required in many bridge circuits is the p.d. across the terminals connected to the detector, or alternatively the current flowing in the detector, although in certain types of bridge the phase of this p.d. or current relative to some standard of phase is what is required. It is also common for the detector p.d. or current to be reduced to zero by manipulation of the values of some of the components of the bridge in which case the bridge is said to be "balanced". The advantage of the latter procedure is that no calibration of the p.d. applied to the bridge is necessary and nonlinear relationship between the parameter being measured and the voltage or current in the detector is of no account.

We are not concerned here with the design of bridge circuits, of which there are many, but with the analysis of the circuit and the determination of the balance conditions. The methods of obtaining these are best shown by worked examples.

Example 1. By the use of Thévenin's theorem, or otherwise, derive an exact expression for the current flowing through the galvanometer in the circuit shown in Fig. 11.1(a). The resistance of the cell may be neglected.

Hence show that, when δR is small compared with R, the galvanometer current is nearly proportioned to δR if the other parameters remain constant.

Using Thévenin's theorem and assuming that the bridge network and cell constitute a source supplying a load consisting of the galvanometer we can proceed as follows:

If the galvanometer were removed, the open-circuit p.d. between A and B would be (Fig. 11.1(b)):

$$E\left(\frac{R+\delta R}{2R+\delta R} - \tfrac{1}{2}\right) = E\delta R/(4R+2\delta R)$$

Remembering that the internal resistance of the cell is zero, the resistance between the points A and B, with the galvanometer removed, can be determined by re-drawing the circuit as in Fig. 11.1(c).

The resistance is thus equal to

$$\frac{R}{2} + \frac{R(R+\delta R)}{2R+\delta R} = \frac{R(4R+3\delta R)}{4R+2\delta R}$$

Hence, by Thévenin's theorem, the current through the galvanometer is given by

$$i = \left(\frac{E\delta R}{4R+2\delta R}\right)\left(\frac{4R^2+3R\delta R+4R6+26\delta R}{4R+2\delta R}\right)$$

$$= \frac{E\delta R}{R(4R+3\delta R)+6(4R+2\delta R)}$$

When δR is small compared with R then i is proportional to δR.

Figure 11.2 shows the circuit of a bridge similar in general form to that shown in Fig. 11.1 but having some reactive components and being supplied from an a.c. source. To find the balance condition—i.e. that no current shall flow in the detector—we must satisfy the condition that the drop of p.d. across AB shall

(a)

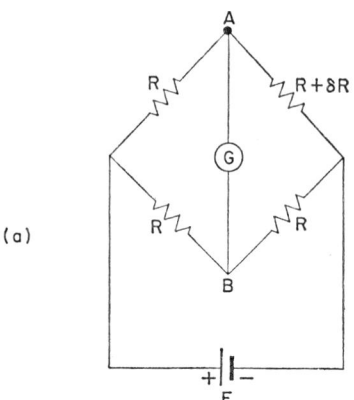

(b)

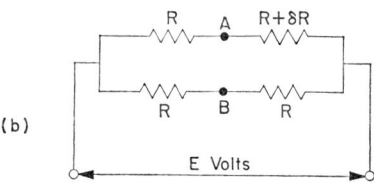

(c)

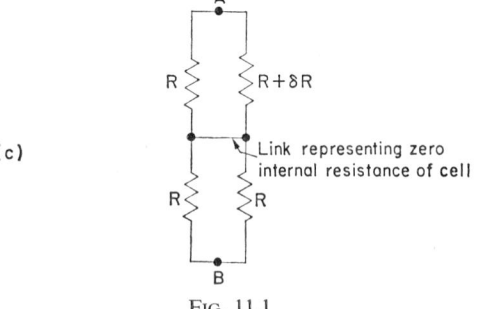

FIG. 11.1

be the same as that across AD and also that these two p.d. shall be in phase so that there is at no instant any net p.d. across BD. These two conditions can usually be obtained by equating real and imaginary parts of the resulting expressions containing j.

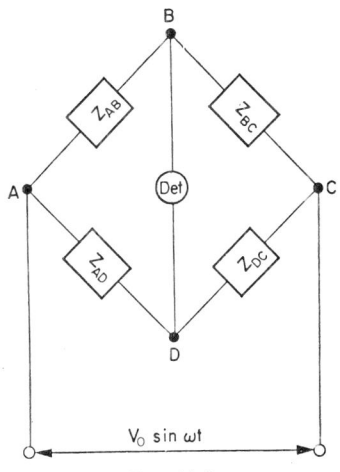

FIG. 11.2

Clearly if $\qquad V_{AB} = V_{AD} \qquad$ then $\qquad V_{BC} = V_{DC}$

Also $\qquad V_{AB} + V_{BC} = V_0 \sin \omega t = V_{AD} + V_{DC}$

so $\qquad \dfrac{V_{AB}}{V_{BC}} = \dfrac{V_{AD}}{V_{DC}} \qquad$ or $\qquad \dfrac{V_{AB}}{V_{AD}} = \dfrac{V_{BC}}{V_{DC}}$

Now, if there is to be no current in the detector, any current flowing in AB must also flow in BC and any current flowing in AD must also flow in DC; hence we can say that, since $V = IZ$:

$$\frac{I_{AB}Z_{AB}}{I_{AB}Z_{BC}} = \frac{I_{AD}Z_{AD}}{I_{AD}Z_{DC}}$$

$$\therefore \frac{Z_{AB}}{Z_{BC}} = \frac{Z_{AD}}{Z_{DC}} \qquad \text{or} \qquad \frac{Z_{AB}}{Z_{AD}} = \frac{Z_{BC}}{Z_{DC}}$$

Example: Find the relations that must exist between the values of the capacitors, the resistors and the angular frequency ω, in the network shown in Fig. 11.3 in order that no current shall flow through the detector.

If $C_1 = 1\,\mu F$, $C_2 = 2\,\mu F$, $R_3 = R_4 = 1000\,\Omega$, $\omega = 1000$ rad/sec, find the phase angle between the currents flowing in R_1 and R_2 respectively when the bridge is balanced.

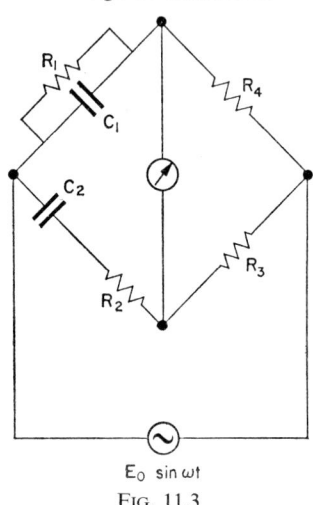

$E_0 \sin \omega t$

FIG. 11.3

Solution: The vector impedance of the branch containing R_1 and C_1 in parallel may be written

$$Z = \left(\frac{-jR_1}{C_1\omega}\right) \Big/ \left(R_1 - \frac{j}{C_1\omega}\right) = -jR_1/(R_1C_1\omega - j)$$

Hence the condition for balance is

$$-jR_1/R_4(R_1C_1\omega - j) = [R_2 - (j/C_2\omega)]/R_3$$

Equating "real" and "imaginary" components

$$R_1R_2C_1C_2\omega^2 = 1$$

$$\frac{C_1}{C_2} = \frac{R_1R_3 - R_2R_4}{R_1R_4}$$

There are, of course, various equivalent pairs of conditions.

When the bridge is balanced, the voltage across R_1 is equal to the voltage across R_2 and C_2 in series. Hence the phase angle between the currents in R_1 and R_2 is:

$$\theta = \tan^{-1}(1/R_2 C_2 \omega)$$

Substitution of numerical values gives

$$\theta = 45°$$

It is possible to design an a.c. bridge which remains in balance whatever the frequency of the supply as is shown by the following example.

Example: Show that the conditions for balance in the bridge circuit of Fig. 11.4 are independent of frequency and find expressions for L and R_1 in terms of the other components.

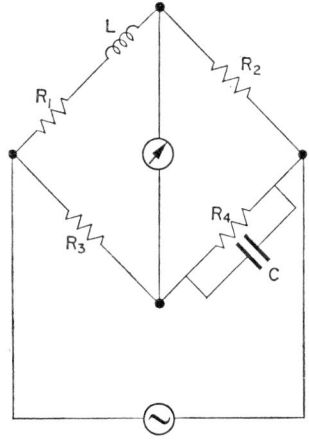

FIG. 11.4

Solution: Balance conditions:

$$\frac{R_1 + j\omega L}{R_3} = R_2 \left/ \frac{R_4}{1 + j\omega C R_4} \right. = \frac{R_2(1 + j\omega C R_4)}{R_4}$$

Equating real and imaginary parts

$$\frac{R_1}{R_3} = \frac{R_2}{R_4} \quad \text{and} \quad \frac{\omega L}{R_3} = \omega C R_2$$

from which ω cancels—hence independence of frequency and

$$R_1 = \frac{R_2 R_3}{R_3}, \quad L = C R_2 R_3$$

An a.c. bridge can also be used to determine the power factor of a practical capacitor—which will always have some loss associated with it—provided that another capacitor, having so little comparative loss that it may be considered ideal, is available.

Example: The bridge shown in Fig. 11.5 is balanced at a frequency of 2·5 kc/s when C_1 has a pure capacitance of 1 μF. Find the capacitance and power factor of the capacitor C_2.

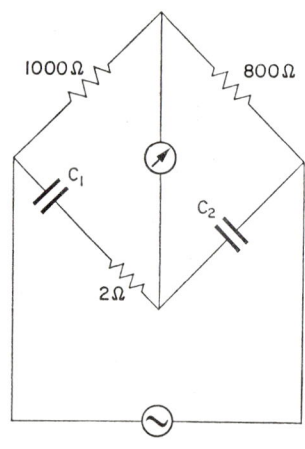

FIG. 11.5

Solution: Let r be the equivalent series resistance of capacitor C_2, then:

$$\frac{1000}{800} = \frac{2 - jX_1}{r - jX_2}$$

$$\therefore \ 1 \cdot 25(r - jX_2) = 2 - jX_1 \ \therefore \ \frac{X_1}{X_2} = \frac{2}{r} = 1 \cdot 25$$

$$\therefore \ C_2 = 1 \cdot 25 \, \mu F, \ r = 1 \cdot 6,$$

$$\therefore \ \text{power factor} = \cos \phi = r\omega C_2 = 0 \cdot 0314$$

Some bridges involve self and mutual inductance as shown in the next example of a frequency-measuring bridge.

Example: Figure 11.6 shows the circuit of a bridge for measuring the frequency of an alternating p.d. applied to terminals A and B.

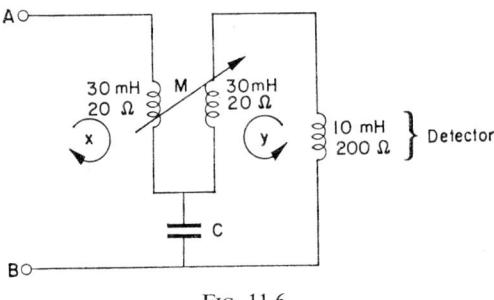

Fig. 11.6

The ideal capacitance C is fixed at $2 \, \mu F$ and the mutual inductance M between the coils is variable from 0 to 20 mH. Determine the relation between M and the frequency for zero current in the detector and calculate the lowest frequency that can be measured with the bridge.

Find also the r.m.s. value of the current in the detector at this same frequency when the mutual inductance is 10 mH and the applied p.d. is 5 V r.m.s.

Solution: Assume circulating currents x and y as shown. For no current in detector

$$j\omega Mx - \frac{jx}{\omega C} = 0$$

$$\therefore \ \omega = \frac{1}{\sqrt{(MC)}}$$

Smallest value of

$$\omega = \frac{1}{\sqrt{(20 \times 10^{-3} \times 2 \times 10^{-6}}} = 5 \times 10^3$$

$$\therefore f = 795 \text{ c/s}$$

When 5 V r.m.s. is applied between A and B, x is given by:

$$5 = x(20 + j150 - j100) + y(j50 - j100)$$

$$\therefore 5 = x(20 + j50) + y(-j50)$$

$$0 = y(200 + j50 + 20 + j150 - j100) + x(j50 - j100)$$

$$\therefore 0 = y(220 + j100) + x(-j50)$$

Eliminating x gives $j250 = y(1900 + j13,000)$

$$\therefore y = \frac{250}{13,130} = 19 \text{ mA}$$

If the e.m.f. induced in coil 1 by current y in coil 2 is ignored, we get an approximate solution as follows:

$$5 = x\left\{20 + j\left(30 \times 10^{-3} \times 5 \times 10^3 - \frac{1}{2 \times 10^{-6} \times 5 \times 10^3}\right)\right\}$$

whence $x = 0.093$ A

Voltage induced in coil 2 $= j \times 10 \times 10^{-3} \times 0.093 = j \times 4.65$

Voltage across capacitor is $-j \times 0.093/(5 \times 10^3 \times 2 \times 10^{-6}) = -j \times 9.3$

$\therefore$ resultant voltage in detector circuit $= -j \times 4.65$

$$\therefore 4.65 = y\sqrt{\left[\left(40 \times 10^{-3} \times 5 \times 10^3 - \frac{1}{2 \times 10^{-6} \times 5 \times 10^3}\right)^2 + 220^2\right]}$$

$$= 242y$$

$$\therefore \quad y = 0.0192 \text{ A or } 19.2 \text{ mA}.$$

Example: Figure 11.7 shows the circuit of an a.c. bridge. L and R are the inductance and resistance of the whole loop shown in heavy lines, including the inductive resistance S; r is a resistor, M_1 and M_2 are the mutual inductances between adjacent coils.

Find the conditions of balance for zero currents in the detector at a frequency $\omega/2\pi$.

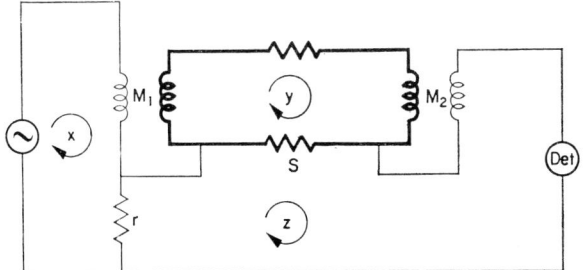

For loop round which current y flows total
self-inductance $= L$, total resistance $= R$.

FIG. 11.7

Solution: Inserting circulating currents x, y and z as shewn and
with the usual convention of sign for M (i.e. self-inductance of
two coils joined in series at their common point is $L_1 + L_2 + 2M$),
if R, L refer to whole of closed loop and $z = 0$ then

For L, R loop $(R + j\omega L)y + j\omega M_1 x = 0$

or $(R + j\omega L)y = -j\omega M_1 x$ (i)

For ($z = 0$) loop $-rx - Sy + j\omega M_2 y = 0$

or $(j\omega M_2 - S)y = rx$ (ii)

    (i) $\times r$ $(Rr + j\omega Lr)y = -j\omega M_1 rx$
    (ii) $\times j\omega M_1$

Adding: $(-\omega^2 M_1 M_2 - j\omega M_1 S)y = j\omega M_1 rx$

$$[(Rr - \omega^2 M_1 M_2) + j\omega(Lr - M_1 S)]y = 0 \quad (y \neq 0)$$

Equating real and imaginary parts:

$$Rr = \omega^2 M_1 M_2$$
$$M_1 S = Lr$$

An elaborate form of this bridge is used by the National
Physical Laboratory to determine the ohm in terms of mutual

inductances M_1 and M_2 which are calibrated from a standard of M calculated from its dimensions and frequency. R is compared with r, a standard coil nominally 20 Ω, by a Wheatstone bridge immediately after the a.c. readings are taken. Hence Rr and R/r being known, r is determined "absolutely".

In many a.c. bridges the impedances being compared are of the same sort, so that their power factors, i.e. the ratio between their series resistance and their series reactance, may be adjusted to be equal by adding series resistance in one arm of the bridge as shown in the example above. Figure 11.8 is an example of such a bridge and the balance conditions are:

$$\frac{X_3}{X_4} = \frac{R_2}{R_4} = \frac{R_1}{R_2} = S$$

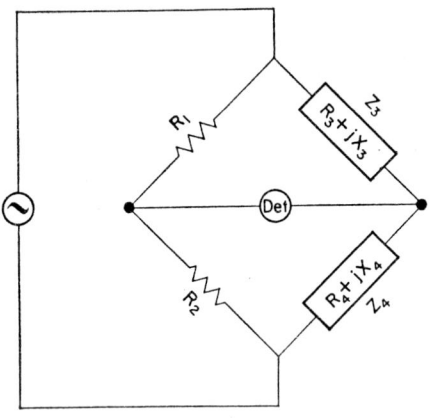

FIG. 11.8

There is, however, great advantage in replacing R_1 and R_2, which act as a potential divider, by two coils mutually coupled as shown in Fig. 11.9. Assuming that L_1 and L_2 have negligible resistance and that $M = \sqrt{(L_1L_2)}$ then:

$$\frac{E_1}{E_2} = \frac{j\omega L_1 I_1 + j\omega M I_2}{j\omega L_2 I_2 + j\omega M I_1}$$

$$= \frac{j\omega L_1 I_1 + j\omega \sqrt{(L_1 L_2)} I_2}{j\omega L_2 I_2 + j\omega \sqrt{(L_1 L_2)} I_1} = \frac{\sqrt{L_1}}{\sqrt{L_2}}$$

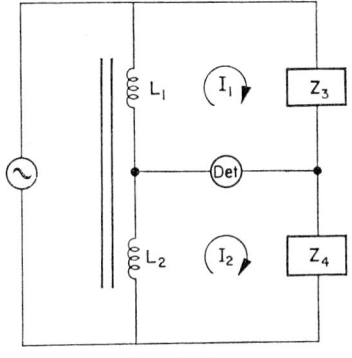

FIG. 11.9

If the coils are wound on a high permeability core so that ideally *all* their flux links *all* their turns then their inductances will be proportioned to (turns)^2 since $L = \phi T/I$ and the flux ϕ is itself proportioned to (current) × (turns) or IT.

Hence we have for balance

$$\frac{Z_3}{Z_4} = \frac{E_1}{E_2} = \frac{N_1}{N_2} = S = \text{turns ratio}$$

Two advantages come from the use of this arrangement. Firstly the ratio E_1/E_2 is independent of the currents flowing in the two coils and therefore of any impedances connected across them. Thus, if we earth the point P in Fig. 11.10, stray capacitances from A and B to earth will be connected across L_1 and L_2 but this will not affect the balance of the bridge. At balance, both P and Q will be at earth potential so stray capacitances from these points to earth will not matter.

The second advantage lies in the precision with which the bridge ratio s is known. L_1 and L_2 will have resistance and the

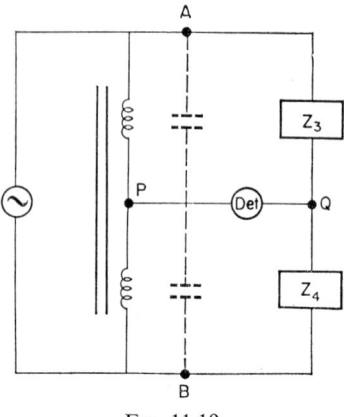

FIG. 11.10

coupling between them cannot be perfect, but inductively coupled ratio arms can be constructed for which s is known with an error less than 1 part in 10^4, which is less than the error in many resistance boxes. If more elaborate precautions are taken with the inductive ratio arms, the error in the value of s need not exceed 1 part in 10^7.

High precision can be attained with values of s up to 100 and with frequencies up to 10 kc/s. Ratio arms can be constructed for higher values of s and for much higher frequencies, but errors are then somewhat larger. Because the coupled inductive ratio arms constitute a transformer, the arrangement is often known as a *transformer bridge*.

Typical Examination Questions

1. The circuit shown in Fig. 1 is used to measure capacities. Show that there will be no voltage across XY if

$$C' = C_1 \frac{R_2}{R_3}, \qquad R' = R_3 \frac{C_2}{C_1}$$

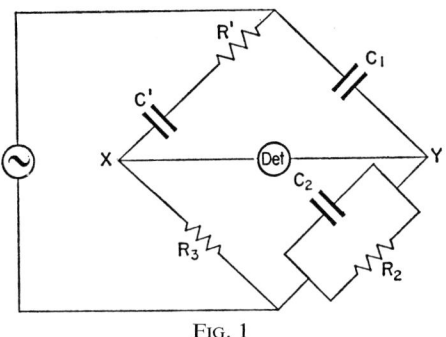

FIG. 1

2. For the circuit of Fig. 2, find the generator current and the potentials of points 1, 2 and 3 with respect to 0 using:

(a) mesh analysis,
(b) nodal analysis,
(c) any method which does not involve the solution of simultaneous equations.

All the resistances are 1 Ω.

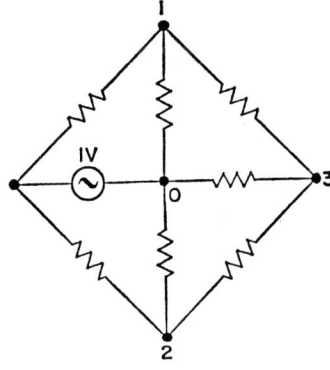

FIG. 2

Ans: $\frac{8}{7}$ A, $\frac{4}{7}$ V, $\frac{2}{7}$ V

3. Derive the balance condition for the a.c. bridge shown in Fig. 3.

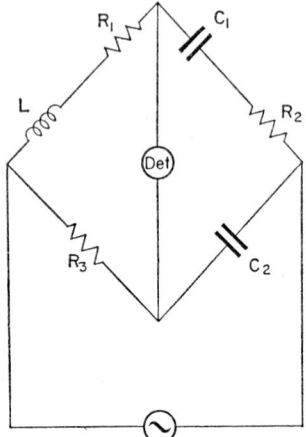

FIG. 3

Ans: $C_1R_1 = C_2R_3$; $L = C_2R_2R_3$

4. Derive the balance condition for the a.c. bridge shown in Fig. 4.

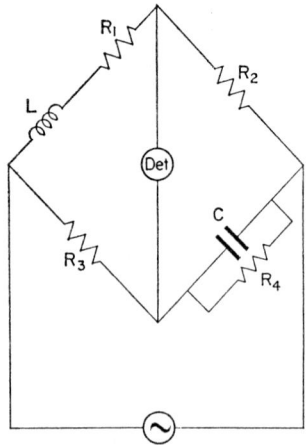

FIG. 4

Ans: $R_1R_4 = R_2R_3 = L/C$

5. Derive the balance condition for the a.c. bridge shown in Fig. 5.

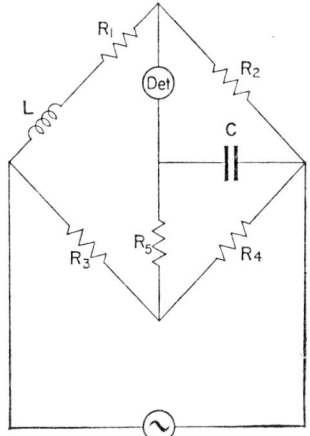

FIG. 5

Ans: $R_1R_4 = R_2R_3$; $LR_4 = CR_2(R_3R_4 + R_3R_5 + R_4R_5)$

CHAPTER 12

Elementary Transmission-line Analysis

EVERY element of length of a transmission line has resistance R, inductance L, leakage conductance G, and capacitance C, to earth and to other conductors. It is usual to assume that these are uniformly distributed, so that at any point each has a value per unit length which is independent of the position of the point.

In practice, a line may not be uniform. Also R is not quite constant owing to temperature and skin effects; and G is dependent on voltage and weather, but the variations are usually small enough for R and G to be taken as constant. L and C can be taken to be constant if the wire spacings are constant and the line is transposed often enough (about once every $1/20$ of a wavelength or more frequently).

The values, per unit length, of the four quantities are the primary constants. They are usually quoted per mile of wire when dealing with power lines, i.e. values for one wire to neutral. Thus for a single-phase circuit the quoted value of:

R will be the resistance per mile of a single conductor,
L will be the inductance to neutral = half that for the wires,
G will be the conductance to neutral = twice that between wires,
C will be the capacitance to neutral = twice that between wires.

Balanced three-phase conditions are worked in terms of one line to neutral (i.e. one leg of an equivalent star). The equivalent single-phase spacing is the geometric mean of the actual three-phase spacings, provided that the line is transposed sufficiently often.

The primary constants can be combined into two terms for any constant frequency f:

1. Series impedance $R + j\omega L = Z$
2. Shunt admittance $G + j\omega C = Y$, where $\omega = 2\pi f$.

Consider an element of a line, of length dl, at a distance l from the receiving end as shown in Fig. 12.1. At the end of the element nearer the receiver let

(i) the voltage from line to neutral $= E$

(ii) and the current in the line $= I$.

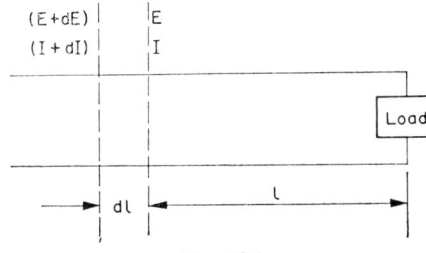

FIG. 12.1

At the other end of the element the voltage can be written $E + dE$ and the current $I + dI$.

The following relations are obtained:

$$dE = IZ \,.\, dl \qquad \text{or} \qquad \frac{dE}{dl} = ZI \qquad (1)$$

$$dI = EY \,.\, dl \qquad \text{or} \qquad \frac{dI}{dl} = YE \qquad (2)$$

Differentiating (1) and (2):

$$\frac{d^2E}{dl^2} = Z\frac{dI}{dl} = ZYE \qquad (3)$$

and

$$\frac{d^2I}{dl^2} = Y\frac{dE}{dl} = ZYI \qquad (4)$$

Solutions of equations (3) and (4) can be written in the forms $E = A \cosh \sqrt{(YZ)}l + B \sinh \sqrt{(YZ)}l$ and $I = A' \cosh (YZ)l + B' \sinh \sqrt{(YZ)}l$, where A, B, A', B' are constants. Solutions can also be written in the forms $E = C e^{\sqrt{(YZ)}l} + D e^{-\sqrt{(YZ)}l}$ and $I = C' e^{\sqrt{(YZ)}l} + D' e^{-\sqrt{(YZ)}l}$, where C, D, C', D' are constants.

At the receiver $l = 0$, $E = E_r$ and $I = I_r$. Thus for the first solution $E = E_r = A$ and from (1)

$$ZI = \frac{dE}{dl} = \sqrt{(YZ)}E_r \sinh \sqrt{(YZ)}l + B \sqrt{(YZ)} \cosh \sqrt{(YZ)}l$$

Thus $ZI_r = B\sqrt{(YZ)}$ and $B = I_r Z_c$ where $Z_c = \sqrt{(Z/Y)}$

Hence $E = E_r \cosh \sqrt{(YZ)}l + I_r Z_c \sinh \sqrt{(YZ)}l$

and $I = I_r \cosh \sqrt{(YZ)}l + \dfrac{E_r}{Z_c} \sinh \sqrt{(YZ)}l$

$Z_c = \sqrt{(Z/Y)}$ is termed the *characteristic impedance* of the line; it is sometimes called the surge impedance or iterative impedance.

$\sqrt{(YZ)}l$ is termed the total angle of the line, while

$\sqrt{(YZ)}$ is the hyperbolic angle per unit length of the line.

The form in exponentials is more useful for long lines than the solution in hyperbolic functions. The exponential solution also shows up the effect of the characteristic impedance, it gives the conditions necessary for resonance, and it gives a method of predicting the voltage and current at any point along the line in terms of either the sender or receiver voltage or current.

At a point l from the receiver $E_l = E'_r e^{\sqrt{(YZ)}l} + E''_r e^{-\sqrt{(YZ)}l}$ for the voltage, and for the current $I_l = I'_r e^{\sqrt{(YZ)}l} + I''_r e^{-\sqrt{(YZ)}l}$ where E'_r, E''_r, I'_r, I''_r are constants.

As $\sqrt{(YZ)}l$ is a complex function, we may write

$$e^{\sqrt{(YZ)}l} = e^{\alpha l + j\beta l} = e^{\alpha l} \cdot e^{j\beta l} = e^{\alpha l} \underline{|\beta l},$$

which is equivalent to a vector of length $e^{\alpha l}$ at an angle βl rad to a reference vector.

Therefore as l increases, i.e. as we consider points along the line at increasing distances from the receiver, $E'_r e^{\sqrt{(YZ)}l}$ represents a voltage which is increasing exponentially and advancing in phase, while $E''_r e^{-\sqrt{(YZ)}l}$ represents a voltage which is decreasing exponentially and retarding in phase. The actual voltage at any point along the line is the vector sum of these two quantities each of which, when regarded for the whole line, is termed a "wave train".

Note that when $l = 0$, $E_r = E'_r + E''_r$ and $I_r = I'_r + I''_r$.

From equation (2) and the above equations on p. 196.

$$\frac{dI}{dl} = YE = I'_r \sqrt{(YZ)} e^{\sqrt{(YZ)}l} - I''_r \sqrt{(YZ)} e^{-\sqrt{(YZ)}l}$$

$$= Y[E'_r . e^{\sqrt{(YZ)}l} + E''_r . e^{-\sqrt{(YZ)}l}]$$

Hence, equating coefficients,

$$\frac{E'_r}{I'_r} = \sqrt{(Z/Y)} = Z_c \quad \text{and} \quad \frac{E''_r}{I''_r} = -Z_c$$

Thus, at any point along the line, the voltage and current vectors of the wave trains are related so that their vector quotients are Z_c for the train which increases as the generator is approached, and $-Z_c$ for the train which increases as the receiver is approached. For both voltage and current there are two trains as indicated below.

At a point l from the receiver, trains increasing with l are

$$E' = E'_r . e^{\alpha l} \underline{|\beta l} \quad \text{and} \quad I'_l = I_r . e^{\alpha l} \underline{|\beta l} \quad \text{where} \quad \frac{E'_l}{I'_l} = Z_c$$

and trains increasing as l decreases are

$$E''_l = E''_r . e^{-\alpha l} \underline{|\beta l} \quad \text{and} \quad I''_l = I''_r . e^{-\alpha l} \underline{|-\beta l}, \quad \text{where} \quad \frac{E''_l}{I''_l} = -Z_c$$

$\sqrt{(YZ)}$ can be written in the form $\alpha + j\beta$, where both α and β are real. α, the real part of $\sqrt{(YZ)}$, is called the attenuation constant as it determines the rate at which the magnitudes of the vectors in the wave trains change exponentially with changing position along the line. Its least value is $\sqrt{(RG)}$, which occurs when the frequency is zero, i.e. for d.c. β, the imaginary part of $\sqrt{(YZ)}$, is called the *wavelength* or *phase-constant* as it determines the rate at which the vectors change their angles with changing position along the line. If l increases by an amount L such that $L = 2\pi$, l must have changed by an amount equal to one complete wavelength. Thus for one wavelength $L = 2\pi/\beta$.

Multiplying the wavelength by the frequency gives the *velocity propagation* of a disturbance; thus the velocity of propagation, or the phase velocity, is $2\pi f/\beta$.

Note. Approximately for air lines $2\pi f/\beta = 186{,}000$ miles/sec. For cable circuits the phase velocity may be as low as 15,000 miles/sec, the wavelength at high frequency being then but a few miles long, a fact which may be important in telephone or telegraph working.

By considering terminal conditions, relations can be obtained between the voltages or currents of the wave trains. By the principle of superposition we have:

Receiver voltage = sum of voltages of wave trains at receiver

Receiver current = sum of currents of wave trains at receiver.

Thus $E_r/I_r = Z_r$ = impedance of load at receiver, and

$$E_r = E_r' + E_r'', \quad I_r = I_r' + I_r'', \quad \text{and} \quad \frac{E_r'}{I_r'} = -\frac{E_r''}{I_r''} = Z_c$$

Therefore $\quad Z_r = \dfrac{E_r' + E_r''}{I_r' + I_r''} \quad$ and $\quad Z_c = \dfrac{E_r' - E_r''}{I_r' + I_r''},$

so that $\quad \dfrac{E_r'}{E_r''} = -\dfrac{I_r'}{I_r''} = \dfrac{Z_r + Z_c}{Z_r - Z_c}, \quad \dfrac{E_r'}{E_r} = \dfrac{Z_r + Z_c}{2Z_r}$

and $\quad \dfrac{E_r''}{E_r} = \dfrac{Z_r - Z_c}{2Z_r}$

Also $\quad \dfrac{I_r'}{I_r} = \dfrac{Z_c + Z_r}{2Z_c} \quad$ and $\quad \dfrac{I_r''}{I_r} = \dfrac{Z_c - Z_r}{2Z_c}$

Receiver open circuited, $Z_r = \infty$.

Therefore $\quad E_r' = \tfrac{1}{2}E_r = E_r'', \quad$ and $\quad I' = \tfrac{1}{2}(E_r/Z_c) = -I_r''.$

Thus, if one wave train is taken to be the reflection of the other,

for voltage: The magnitudes of the received and reflected trains are equal and the phases are the same;

for current: The magnitudes of the received and reflected trains are equal but the phases are reversed.

The expressions given above for voltage and current at any point l from the receiver now become

$$E = \tfrac{1}{2}E_r[e^{\sqrt{(YZ)}\,l} + e^{-\sqrt{(YZ)}\,l}] = E_r\cosh\sqrt{(YZ)}l$$

(cf. putting $I_r = 0$ in the equations on p. 196)

$$I = \frac{1}{2}\frac{E_r}{Z_c}[e^{\sqrt{(YZ)}\,l} - e^{-\sqrt{(YZ)}\,l}] = \frac{E_r}{Z_c}\sinh\sqrt{(XY)}l$$

If the receiver is short-circuited there is a reversal of phase of the voltage.

$$Z_r = 0, \qquad I'_r = \tfrac{1}{2}I_r = I''_r, \qquad \text{and} \qquad E'_r = \tfrac{1}{2}Z_cI_r = -E''_r.$$

Hence $\qquad I = \tfrac{1}{2}I_r[e^{\sqrt{(YZ)}l} + e^{-\sqrt{(YZ)}l}] = I_r\cosh\sqrt{(YZ)}l$

(cf. putting $E_r = 0$ in equations on p. 196)

$$E = \tfrac{1}{2}Z_cI_r[e^{\sqrt{(YZ)}l} - e^{-\sqrt{(YZ)}l}] = Z_cI_r\sinh\sqrt{(YZ)}l.$$

For load impedance equal to Z_c there is no reflection, for

$$E''_r = 0 = I''_r, \qquad E'_r = E_r \qquad \text{and} \qquad I'_r = I_r$$

Thus $\qquad E = E_r e^{\sqrt{(YZ)}l} \qquad$ and $\qquad I = I_r e^{\sqrt{(YZ)}l}.$

The wave train theory gives an explanation of the formation of the resonance peaks and troughs which occur in the voltage and current distributions along transmission lines. When the two voltage or current vectors of their respective trains are nearly in-phase or anti-phase a peak or trough will occur, the magnitude of the peak or trough depending on the attenuation which may have occurred, i.e. it is dependent on the real part of $\sqrt{(YZ)}$.

As the vectors of a pair of trains rotate in opposite directions, the interval between successive resonance points will be approximately equal to the distance in which one vector rotates through $\pi/2$, i.e. a distance equal to a quarter-wavelength. The interval between two successive peaks will be a half-wavelength. The distance of the first peak or trough from the receiver will depend on the angle of the load impedance.

If the receiver is open-circuited then, at the receiver, the voltage vectors are in-phase. After a quarter-wavelength, each will have rotated $\pi/2$ so that they will be in anti-phase and a resonance trough will occur. After a further quarter-wavelength the vectors will be in-phase again and there will be a resonance peak. Similar troughs and peaks will occur at intervals of approximately a quarter-wavelength, if the attenuation constant is small, which it usually is. As will be shown, the exact position of the resonance points depend on both the attenuation and the wavelength constants.

At the receiver, the current vectors are opposing, which corresponds to a trough, thus there will be a peak at a quarter-wavelength from the receiver, and a trough at a half-wavelength.

When the receiver is short-circuited the above conditions are reversed, the first resonance point from the receiver being a peak for voltage and a trough for current.

When the load has the value of the characteristic impedance there is no reflected wave train and therefore no resonance points. The voltage and current distributions are simple exponentials and are the same as would be obtained if the line were of infinite length. As the frequency applied to any line is increased, the wavelength decreases; thus a greater number of resonance points may occur on the line. Increasing the frequency is electrically equivalent to lengthening the line. Curves of voltage or current distribution plotted against wavelength, instead of line length, would be independent of frequency.

When α is small, and with an open or short-circuited receiver, at resonance βl is a multiple of $\pi/2$. Thus for an open receiver, if l is an even multiple of a quarter-wavelength, βl is an even multiple of $\pi/2$ and thus:

$$E_l = \tfrac{1}{2}E_r(e^{\alpha l} + e^{-\alpha l}) = E_r \cosh \alpha l \qquad \text{(peak)}$$

$$I_l = \frac{1}{2Z_c} E_r(e^{\alpha l} - e^{-\alpha l}) = \frac{E_r}{Z_c} \sinh \alpha l \qquad \text{(trough)}$$

If l is an odd multiple of a quarter-wavelength,

$$E_l = \tfrac{1}{2}E_r(e^{\alpha l} - e^{-\alpha l}) = E_r \sinh \alpha l \qquad \text{(trough)}$$

$$I_l = \frac{1}{2Z_c} E_r(e^{\alpha l} + e^{-\alpha l}) = \frac{E_r}{Z_c} \cosh \alpha l \qquad \text{(peak)}$$

For a shorted receiver, if l is an even multiple of a quarter-wavelength

$$I_l = \tfrac{1}{2}I_r(e^{\alpha l} + e^{-\alpha l}) = I_r \cosh \alpha l \qquad \text{(peak)}$$

$$E_l = \frac{Z_c I_r}{2}(e^{\alpha l} - e^{-\alpha l}) = Z_c I_r \sinh \alpha l \qquad \text{(trough)}$$

If l is an odd multiple of a quarter-wavelength

$$I_l = \tfrac{1}{2}I_r(e^{\alpha l} - e^{-\alpha l}) = I_r \sinh \alpha l \qquad \text{(trough)}$$

$$E_l = \frac{Z_c I_r}{2}(e^{\alpha l} + e^{-\alpha l}) = Z_c I_r \cosh \alpha l \qquad \text{(peak)}$$

With a load which is greater than the characteristic impedance, voltage and current distributions will be obtained which are similar to those for an open receiver, but the amplitudes of the standing waves will be smaller. This is due to the fact that the reflected wave has a smaller value than in the open-circuited case. Similarly, in the case of a load which is smaller than the characteristic impedance, the voltage and current distributions will be similar to those in the short-circuited case, but the peaks will be smaller. The positions and sizes of the resonance peaks and troughs depend on the size and phase-angle of the load impedance.

Exact equations can be obtained for the positions of resonance points, but, as will be seen, solutions of the equations can only be obtained in particular cases.

Let $Z_c = Z_r \cdot A(\cos \theta + j \sin \theta)$, where Z_r is the load impedance and A and θ are real. Then the hyperbolic solution gives

$$E_l = E_r[\cosh{(\alpha l + j\beta l)} + (Z_c/Z_r)\sinh{(\alpha l + j\beta l)}]$$
$$= E_r[\cosh{\alpha l}\cos{\beta l} + j\sinh{\alpha l}\sin{\beta l}$$
$$+ A(\cos{\theta} + j\sin{\theta})(\sinh{\alpha l}\cos{\beta l} + j\cosh{\alpha l}\sin{\beta l})]$$

Separating the real and imaginary parts, squaring and adding, the expression for the square of the magnitude of E_l/E_r reduces to

$$\cosh^2{\alpha l}(A^2\sin^2{\beta l} + \cos^2{\alpha l}) + \sinh^2{\alpha l}(A^2\cos^2{\beta l} + \sin^2{\beta l})$$
$$- A\sin{\theta}\sin{2\beta l} + A\cos{\theta}\sinh{2\alpha l}$$

Differentiating with respect to l for a maximum or minimum we have

$$(A^2 + 1)\alpha\sinh{2\alpha l} + (A^2 - 1)\beta\sin{2\beta l} - 2A\beta\sin{\theta}\cos{2\beta l}$$
$$+ 2A\alpha\cos{\theta}\cosh{2\alpha l} = 0$$

Similarly for the current maxima and minima we have

$$(1 + A^2)\alpha\sinh{2\alpha l} + (1 - A^2)\beta\sin{2\beta l} + 2A\beta\sin{\theta}\cos{2\beta l}$$
$$+ 2A\alpha\cos{\theta}\cosh{2\alpha l} = 0$$

When $Z_r = Z_c$, $A = 1$ and $\theta = 0$ the terms involving βl disappear and there are no resonance points.

On open-circuit load, $A = 0$, and for resonance points of voltage

$$\alpha\sinh{2\alpha l} - \beta\sin{2\beta l} = 0$$

while for those of current

$$\alpha\sinh{2\alpha l} + \beta\sin{2\beta l} = 0$$

On short-circuit load, $A = \infty$, and for resonance points of voltage

$$\alpha\sinh{2\alpha l} + \beta\sin{2\beta l} = 0$$

while for those of current

$$\alpha\sinh{2\alpha l} - \beta\sin{2\beta l} = 0$$

If l is small, so that $\sinh{2\alpha l}$ can be taken to be zero, the above relations reduce to $\sin{2\beta l} = 0$. In this case the positions of the resonance points will be given by $\beta l = n\pi/2$, where n is an integer, a result which is in accordance with those obtained from the wave train theory.

Typical Examination Questions

1. A line has inductance $1 \cdot 25 \times 10^{-6}$ H/m and capacitors $2 \cdot 0 \times 10^{-11}$ F/m. Losses are negligible. Find (i) the characteristic impedance and admittance, (ii) the time taken for a voltage pulse to travel 25 m, (iii) the wavelength along the line if it is excited by sinusoidal a.c. at 25 Mc/s.

Ans: 250 Ω; $4 \cdot 10^{-3}$ Ω^{-1}; $0 \cdot 125$ μs, 8 m

2. If the line in Qu. 1 is terminated at a length of 25 m by a resistance of 375 Ω; and a unit step-voltage is applied at the sending end by a generator of negligible internal impedance, what are the amplitudes of (i) the first reflected voltage wave from the 375 Ω, (ii) the first reflected voltage wave from the generator? Hence sketch the waveform of the voltage produced across the 375 Ω.

Ans: $0 \cdot 2$, $-0 \cdot 2$; step to $1 \cdot 2$ after $0 \cdot 125$ μs, followed by steps each $0 \cdot 25$ μs to $1 - 0 \cdot 2^2$, $1 + 0 \cdot 2^3$, $1 - 0 \cdot 2^4$, etc.

3. If the excitation in Qu. 2 was sinusoidal a.c. at 25 Mc/s instead of a step, find (i) the reflection coefficient at the 375 Ω, (ii) the V.S.W.R., and (iii) the reflection coefficient half-way along the line.

Ans: $0 \cdot 2 \lfloor 0$; $1 \cdot 5$; $0 \cdot 2 \lfloor -45$

4. What is the input impedance, seen by the source, in Qu. 3?

Ans: 250 $\lfloor 22 \cdot 6°$

5. A $1/4$ λ section of line can be used between the end of the line in Qu. 2 and the 375 Ω load, to remove standing waves. What should the characteristic admittance of this section be?

Ans: $3 \cdot 26 \times 10^{-3}$ Ω^{-1}

6. The load can also be matched by a stub, short-circuited at its end, connected across the line. Show that this should be connected $0 \cdot 141$ λ from the 375 Ω, and find its length if it has $Y_o = 4 \times 10^{-3}$ Ω^{-1}.

Ans: $0 \cdot 188$ λ

7. If the same line is terminated by an admittance $(1 + j)$ 4×10^{-3} Ω^{-1} instead of the 375 Ω, find (i) the V.S.W.R., (ii) the admittance at the input of the line.

Ans: $2 \cdot 62$, $(2 - j) \times 4 \cdot 10^{-3}$ Ω^{-1}

CHAPTER 13

Tuned Circuits and Filters

A TUNED electrical network responds differently to different frequencies and in this sense the a.c. bridge networks discussed in chapter 11 are tuned. Similarly an electrical filter is a network which offers less impedance to currents of certain frequencies than to those at other frequencies and so a filter is also a tuned network. The relationship between any chosen parameters for the network is known as its response curve or characteristic and it is usually the aim of the designer of a network to achieve a characteristic as close to an ideal one as is possible with reasonable economy. The desired characteristic may be of any shape but is very frequently rectangular. Thus an ideal for a tuning network in a radio receiver might be that its impedance should be infinite over the frequency range 1 Mc $\pm$ 5 kc/s and zero for all other frequencies. Because the components making up the network are not ideal and invariably have unwanted losses associated with them it is not usually possible to obtain a perfect response curve.

The purpose of the circuits we shall now consider is the separation of currents or voltages of one frequency from those of another, so the frequency characteristics of the components available for building these circuits are of prime importance. Such characteristics for resistors, inductors and capacitors and also for series and parallel tuned circuits are illustrated graphically in Fig. 13.1, where, in each case, impedance is plotted against frequency.

Although the resistance of a resistor does not change with frequency, it can often be used in place of an inductor when the

204

Circuit   Response

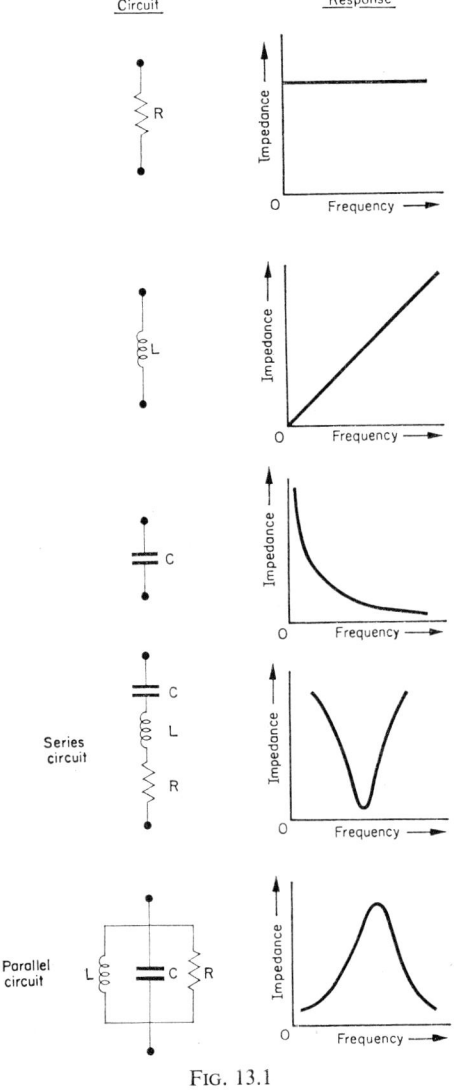

Series
circuit

Parallel
circuit

FIG. 13.1

latter would be difficult and expensive to construct with sufficiently large values of reactance.

Series and parallel tuned circuits are in several instances ideal components for simple filters since they provide impedances which change rapidly with frequency. However, rapid changes of impedance are obtained only when the circuits have high ratio of reactance to resistance, known as the *Q factor*, and the greater part of the change then occurs in a narrow band of frequencies. Such circuits are therefore very valuable for the segregation of a narrow band of frequencies and the theory relating to their use for this purpose is dealt with later in this chapter. They are much less useful when relatively wide bands of frequencies have to be passed through the filter.

In many tuned circuits and filter circuits we are concerned with the flow of current from the source to the load. We wish to connect between source and load additional elements, which will hinder the flow of current to the load at one frequency while leaving it relatively unchanged at another. Using Thévenin's theorem, the source and load can be represented by the equivalent circuit of Fig. 13.2. For simplicity the source and load impedances

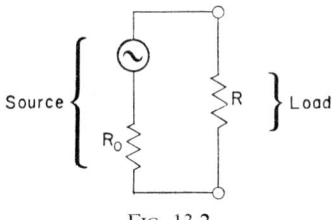

Fig. 13.2

are shown as pure resistances, since this is a close approximation to the truth in many cases. If reactive components were present, what follows would not be greatly affected, though the possibility of unwanted peaks in the response curve occurring would have to be taken into account when adding other components.

If we add only one element of impedance Z, the resulting circuit must take the form of Fig. 13.3(a) or Fig. 13.3(b).

In Fig.13.3(a) Z acts by reducing the flow of current at the unwanted frequencies while leaving it relatively unchanged at the wanted frequencies. Thus if Z were an inductor it would attenuate high frequencies in comparison with low, whereas a capacitor

(a)

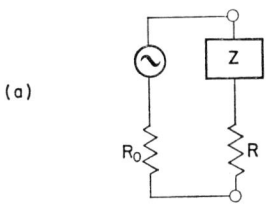

(b)

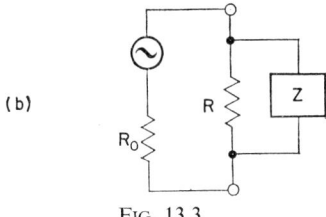

FIG. 13.3

would have the reverse effect. A parallel resonant circuit would attenuate a narrow band of frequencies in the neighbourhood of the resonant frequency.

It is most important to realize that the addition of Z will have little effect unless the impedance of Z at the unwanted frequency is large compared with both R_0 and R. Furthermore, in circuits of this kind we are generally seeking a large attenuation of the unwanted current—say by a factor of ten or more. Hence the addition of Z as shown in Fig. 13.3(a) is of little use unless we can arrange for Z to be at least ten times as great as R_0 or R_1 at the unwanted frequency.

In Fig. 13.3(b), the function of Z is to by-pass, at the unwanted frequency, current which would otherwise flow through R. Thus if Z were an inductor it would attentuate low-frequency currents in the load R in comparison with high frequencies, while a capacitor would have the reverse effect. A series-resonant circuit would attenuate a band of frequencies in the neighbourhood of the resonant frequency.

Clearly the by-passing effect of Z will be small unless the value of Z at the unwanted frequencies is small in comparison with R—less than one-tenth of R, say. It is less obvious that Z must also be small compared with R_0, but if R_0 were negligible, the only effect of adding Z would be to draw additional current from the generator; the current through the load would have the constant value E/R. The circuit operates efficiently only if the additional current through Z causes a large voltage drop across R_0 and thus reduces the current through R at the unwanted frequency. For this to happen, Z must be small compared with R_0.

We have seen that the circuit arrangement of Fig. 13.3(a) operates satisfactorily if R and R_0 are both fairly low while that of Fig. 13.3(b) works well when R and R_o are both fairly large. Cases arise, however, where R may be large and R_0 small, or vice versa. When this happens, the addition of a single impedance Z cannot normally provide satisfactory filtering, so we must consider the addition of two impedances.

If R_0 is high and R low, the arrangement should obviously be that shown in Fig. 13.4(a).

If we wish to attenuate high frequencies in comparison with low, the circuit of Fig. 13.4(b) would be correct, while the circuit of Fig. 13.4(c) would have the opposite effect.

On the other hand, if R_0 is low and R high, the correct circuit is shown in Fig. 13.5(a).

The version shown in Fig. 13.5(b) would attenuate high frequencies, while low frequencies would be attenuated by the circuit of Fig. 13.5(c).

The criterion for choice of value of R and R_0 must be whether R or R_0 is high or low in comparison with values that are given to Z_1 and Z_2.

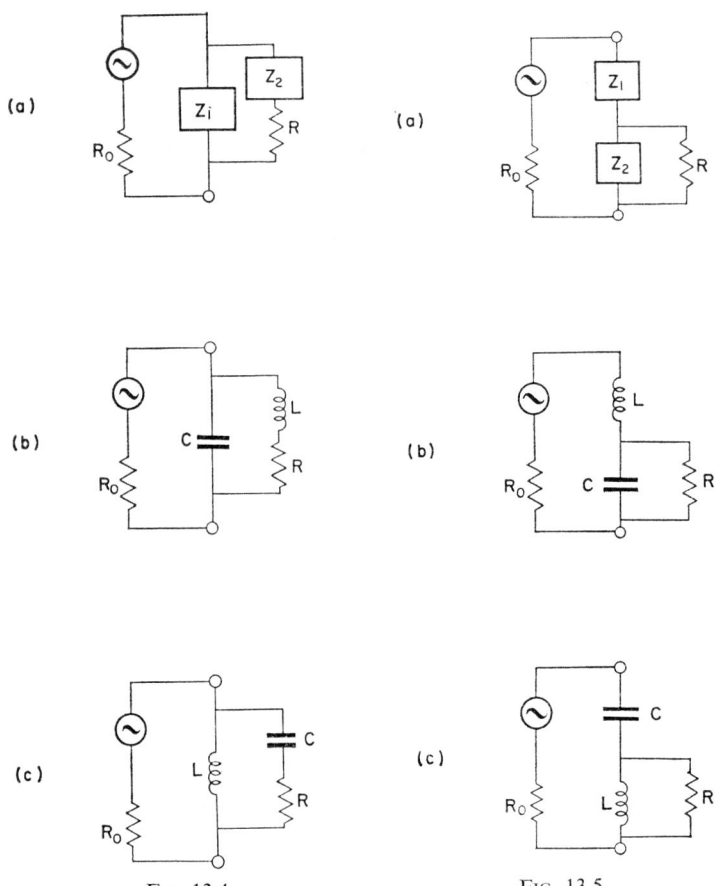

(a)

(a)

(b)

(b)

(c)

(c)

FIG. 13.4

FIG. 13.5

A simple filter of the type discussed above is shown in Fig. 13.6. It is to be used for the reduction to one-tenth of its original value of the residual ripple in the output of a rectifier system which is giving a mean output of 100 V with a superimposed sinusoidal ripple of ± 10 V peak at a frequency of 100 c/s. It is required to find suitable values for L and C assumed to be ideal.

The operation of the filter can best be understood if its effect on the direct and alternating components are considered separately. If the inductor is ideal it will have no resistance, and hence the direct component of current will flow unimpeded to the load, and none will pass through the capacitor. The alternating component driven by the alternating p.d. constituting the ripple will be impeded by the reactive components and will flow

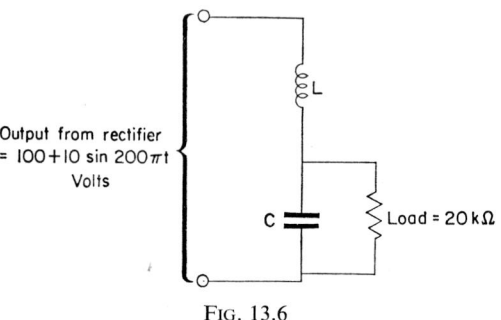

Output from rectifier = $100 + 10 \sin 200\pi t$ Volts

Load = 20 kΩ

FIG. 13.6

through the capacitor and the load. If we make the capacitor such that its reactance at 100 c/s is much less than the resistance of the load, then almost all the alternating component of current will flow through it and we may neglect the effect of the load shunting it. Then we may say that the impedance of the circuit to alternating circuits is purely reactive and is:

$$Z_{AC} = j\omega L - \frac{j}{\omega C}$$

$$\therefore I_{AC} = \frac{V_{AC}}{Z_{AC}} = \frac{V_{AC}}{(j\omega L - j/\omega C)}$$

$$\therefore \text{ripple across } C = I_{AC} \times \frac{-j}{\omega C} = V_{AC} \frac{-j/\omega C}{(j\omega L - j/\omega C)}$$

Thus the ripple voltage has been reduced in the ratio

$$\frac{-j/\omega C}{(j\omega L - j/\omega C)}$$

This ratio is to be $1/10$,

$$\therefore 10 \times \frac{-j}{\omega C} = j\omega L - \frac{j}{\omega C}$$

$$\therefore j\omega L = -11 \frac{j}{\omega C}$$

We must now choose a value of C such that $1/\omega C$ is small compared with 20 kΩ. Let C be $10 \mu\text{F}$ then:

$$X_c = \frac{1}{\omega C} = \frac{10^6}{200\pi \times 10} = 159 \,\Omega$$

This is less than one-hundredth of the load resistor and will be quite satisfactory and thus:

$$\omega L = 11 X_c = 1749 \,\Omega$$

$$\therefore L = \frac{1749}{200\pi} = 27 \cdot 8 \text{ H}$$

If further reduction of ripple is desired another similar network can be added to that shown in Fig. 13.6.

It will be seen from the above analysis that so far as a.c. components are concerned the network behaves as an a.c. potential divider and is drawn in the form shown in Fig. 13.6 to help in appreciating this point.

We shall next discuss simplified calculations on resonant circuits, which are valid when the Q, as defined above, is sufficiently high, that is above say 20. Tuned circuits which obey this condition are very commonly encountered.

The tuned circuit is built up from inductors which are not perfect, from capacitors which are not perfect and from resistance which may be partly due to the effects of external loads. As a preliminary we shall consider the behaviour of practical inductors and capacitors alone.

Inductors may be either air-cored or wound on metal or ferrite cores. When a.c. flows through the inductor power losses arise due to several causes, the more important being:

(a) The a.c. resistance of the winding. This is different from the d.c. resistance because of skin effect and therefore is frequency dependent.

(b) Eddy-currents induced in the winding and cores introduce loss.

(c) Resistance loss through insulation.

(d) Losses due to coupling with other circuits.

All these sources of loss depend on frequency in different ways so that it is impossible to represent the real inductance by an equivalent circuit in a way which is valid for all frequencies. However, for a single frequency or for a small band of frequencies a very simple equivalent can always be drawn, as in Fig. 13.7

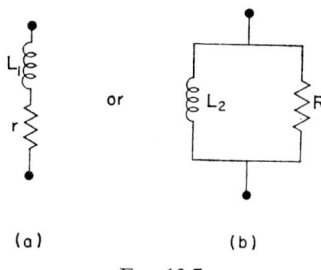

(a)   (b)

Fig. 13.7

Since the loss components will reduce the lagging phase angle from the ideal value of 90°, either of the equivalents (a) or (b) can be used.

In capacitors the power losses arise from:

(a) leakage through the dielectric,

(b) leakage over the surfaces and between terminals,

(c) the series resistance of the plates, and from other minor causes.

Once more a single frequency equivalent is easily found and may be as shown in Fig. 13.8(a) and (b).

In a good inductor r should be much less than the reactance ωL and therefore $\omega L/r$ is a useful measure of the efficiency of the inductor.

$$Q_L = \frac{\omega L_1}{r}$$

For the capacitor, $\qquad Q_c = \dfrac{1}{\omega C_1 r}$

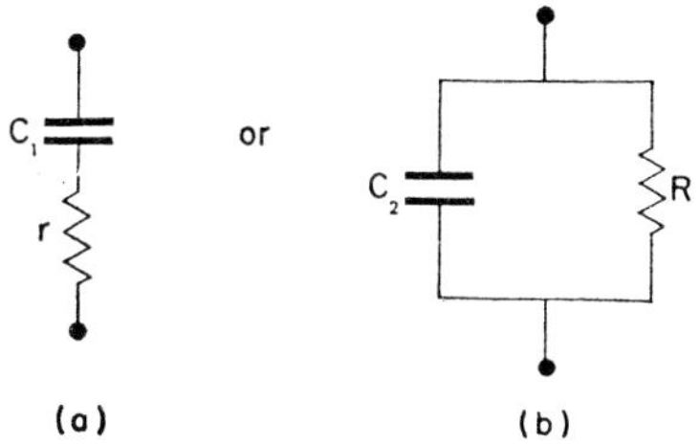

(a)   (b)

FIG. 13.8

At audio and radio frequencies, $Q_L = 10–300$ for wound inductors and at audio and radio frequencies $Q_c > 100$ for normal capacitors. The relations between the equivalent circuits (a) and (b), Fig. 13.7 may be determined as follows:

Equate admittances giving:

$$\frac{1}{r + j\omega L_1} = \frac{1}{R} - \frac{j}{\omega L_2}$$

but $\qquad \dfrac{1}{r + j\omega L_1} = \dfrac{r - j\omega L_1}{r^2 + \omega^2 L_1^2} = \dfrac{r}{r^2 + \omega^2 L_1^2} - \dfrac{j\omega L_1}{r^2 + \omega^2 L_1^2}$

Now $\qquad \omega^2 L_1^2 \gg r^2 \text{ for } Q > 20$

$$\therefore \quad \frac{1}{r + j\omega L_1} \simeq \frac{r}{\omega^2 L_1^2} - \frac{j}{\omega L_1}$$

$$= \frac{1}{R} - \frac{j}{\omega L_2}$$

Equating real and imaginary parts:

$$R = \frac{\omega^2 L_1^2}{r} = \omega L_1 Q = Q^2 r$$

$$\frac{\omega L_2}{j} = \frac{\omega L_1}{j} \text{ giving } L_2 = L_1$$

If $Q = 10$, $\omega L_1 = 10r$

$\therefore R = \dfrac{r^2 + 100r^2}{r} = 101r$, instead of $100r$ by the approximate expression.

$$L_2 = \frac{r^2 + \omega^2 L_1^2}{\omega^2 L_1} = L_1[1 + (1/Q^2)] \rightarrow L_1 \text{ when } Q \text{ is large.}$$

Notice that $R = \omega^2 L_1 L_2$ or $\dfrac{R}{\omega L_2} = \dfrac{\omega L_1}{r} = Q$ exactly.

Similarly, for C, where the high Q approximation can always be used in practice:

$$R \doteqdot Q^2 r = \frac{Q}{\omega C_1}$$

$$C_1 = C_2$$

We shall next consider the combination of imperfect circuit elements in a resonant circuit as shown in Fig. 13.9 and Fig. 13.10.

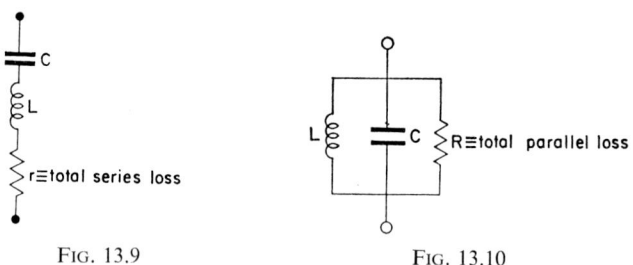

FIG. 13.9   FIG. 13.10

If the circuit is a *series* resonant one we should compound the *series* equivalent circuits, so that total loss $= r_L + r_L$.

If it is a *parallel* resonant one we compound the *parallel* equivalents.

In general we have to define what criterion for resonance is being used. This might be either that i and therefore V_r are max or that V_C is max. (This latter is spoken of as voltage resonance.) Similar considerations apply to the parallel resonant circuit, but, if Q is high then the conditions become identical.

At resonance, therefore, the inductive and capacitative reactances can be taken equal (current resonance in a series circuit) and Q_0, the Q of the circuit, is related to ω_0, the resonant frequency, by

$$Q_0 = \frac{\omega_0 L}{r} = \frac{R}{\omega_0 L} = \frac{1}{\omega_0 C} = \omega_0 CR$$

The band of frequencies passed by a high Q resonant circuit is small so all the phenomena of interest occur close to ω_0 and when we speak of the Q of a resonant circuit we mean the value measured at ω_0, which is considered as a constant over the pass-band of the circuit.

In radio circuits we are nearly always interested in the voltage across L or C. For the series L–C circuit, Fig. 13.9:

At resonance $\qquad i = \dfrac{E}{r} \qquad \therefore\ V_L = j\,\dfrac{E \cdot \omega_0 L}{r} = jEQ$

The voltage measured will differ slightly from this both in magnitude and phase as it will be the voltage across L plus the voltage across part of r, but the above result is sufficiently good for engineering purposes.

For the parallel L–C circuit, Fig. 13.10:

The impedance at resonance is the pure resistance R and the voltage across both L and C is given by

$$V_L = V_C = iR = iQ^2 r$$

We shall next consider the effect of small variations of frequency in the region of resonant frequency ω_0.

(1) *Series circuit.* Let $\omega = \omega_0 + \delta\omega$, $\delta\omega \ll \omega_0$

$$I = \frac{E}{\sqrt{\{r^2 + [\omega L - (1/\omega C)]^2\}}}$$

But
$$\omega L - \frac{1}{\omega C} = L(\omega_0 + \delta\omega) - \frac{1}{C(\omega_0 + \delta\omega)}$$

for
$$\delta\omega \ll \omega_0 \simeq L(\omega_0 + \delta\omega) - \frac{1}{\omega_0 C}\left(1 - \frac{\delta\omega}{\omega_0}\right)$$

$$= \omega_0 L + L\delta\omega - \frac{1}{\omega_0 C} + \frac{\delta\omega}{\omega_0^2 C}$$

Since
$$\omega_0 L = \frac{1}{\omega_0 C} \qquad \therefore \ \omega_0^2 = \frac{1}{LC}$$

$$\therefore \ \omega L - \frac{1}{\omega C} = ZL\delta\omega$$

$$\therefore \ I = \frac{E}{\sqrt{(r^2 + 4L^2\delta\omega^2)}} = \frac{E}{r\sqrt{(1 + 4L^2\delta\omega^2/r^2)}}$$

$$= \frac{E}{r\sqrt{(1 + 4Q^2\delta\omega_0^2/\omega^2)}}$$

$$\frac{1}{C(\omega_0 + \delta\omega)} = \frac{(\omega_0 - \delta\omega)}{C(\omega_0 + \delta\omega)(\omega_0 - \delta\omega)}$$

$$= \frac{\omega_0 - \delta\omega}{C(\omega_0^2 - \delta\omega^2)}$$

$$\simeq \frac{\omega_0 - \delta\omega}{C\omega_0^2}$$

$$= \frac{1}{\omega_0 C}\left(\frac{\omega_0 - \delta\omega}{\omega_0}\right)$$

$$= \frac{1}{\omega_0 C}\left(1 - \frac{\delta\omega}{\omega_0}\right)$$

If $4Q^2\delta\omega^2 = \omega_0^2$, the voltage will have dropped from E/r to $E/(r\sqrt{2})$, approximately $0\cdot 71\ E_{\max}$ or power to half.

$$\therefore \ Q = \frac{\omega_0}{2\delta\omega}$$

where $\delta\omega = \frac{1}{2}$ frequency difference between $0\cdot 71$ voltage points. This is a useful technique for measuring Q.

(2) *Parallel circuit.* Here the admittance Y is

$$Y = \frac{1}{R} - j\left(\frac{1}{\omega L} - \omega C\right)$$

$$= \frac{1}{R} - (j/\omega_0 L)\left(\frac{\omega_0}{\omega} - \frac{\omega}{\omega_0}\right) \qquad \left(\text{because } \omega_0 L = \frac{1}{\omega_0 C}\right)$$

$$= \frac{1}{R} - (j/\omega_0 L)\left(\frac{\omega_0}{\omega_0 + \delta\omega} - \frac{\omega_0 + \delta\omega}{\omega_0}\right) \qquad \left(\therefore C = \frac{1}{\omega_0^2 L}\right)$$

$$= \frac{1}{R} - (j/\omega_0 L)\left(-\frac{2\delta\omega}{\omega_0}\right)$$

$$\left(\text{because } \frac{\omega_0}{\omega_0 + \delta\omega} \simeq \frac{\omega_0 - \delta\omega}{\omega_0}\right)$$

$$= \frac{1}{R} + 2jC\delta\omega$$

$$|Y| = \frac{1}{R}\sqrt{[1 + (2CR\delta\omega)^2]}$$

Now

$$E = \frac{i}{|Y|} = \frac{iR}{\sqrt{[1 + (2CR\delta\omega)^2]}} = \frac{iR}{\sqrt{[1 + (2\delta\omega Q/\omega_0)^2]}}$$

because $\qquad\qquad CR\omega_0 = Q \qquad CR = \dfrac{Q}{\omega_0}$

Once again E is reduced to $0.71\, E_{max}$ when

$$\frac{\omega_0}{2\delta\omega} = Q$$

N.B. When $\delta\omega$ is negative, i.e. below resonance,

$$Y = \frac{1}{R} - 2jC\delta\omega$$

so that the susceptance changes sign in going through resonance.

Example: A parallel tuned circuit has a Q of 80 and a resonant frequency of 100 kc/s. If it has an impedance of 10 kΩ at resonance, estimate the inductance of the coil. This coil is to be used in a circuit tuned to a frequency of 125 kc/s. Calculate the capacitance of the capacitor which must be used to do this.

Solution: The equivalent circuit for the combination is as shown in Fig. 13.10.

$$R = 10^4, \qquad Q = 80 = \frac{R}{\omega L}$$

$$\therefore \ \omega L = \frac{10^4}{80} = 125$$

$$\therefore \ L = \frac{10^4}{80 \times 2\pi \times 10^5} = 199 \ \mu\text{H}.$$

At 125 kc/s:

$$X_c = -125 \times 1 \cdot 25 \ \Omega$$

$$\therefore \ C = \frac{10^{12}}{1 \cdot 25 \times 2\pi \times 10^5 \times 125 \times 1 \cdot 25} = 8160 \ \text{pF}$$

Example: For the circuit given in Fig. 13.11 V is the p.d. across the capacitor and $\delta\omega$ is the amount by which the angular frequency

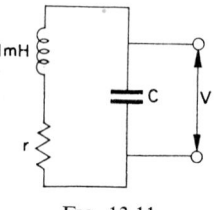

ImH

C

V

r

Fig. 13.11

of currents, induced in the circuit by an outside source producing a constant e.m.f. in the inductor, differ from the resonant angular frequency, 8×10^5 rad/sec. The relation between $\delta\omega$ and V is:

$\delta\omega$	0	1000	2000	3000	4000	5000	6000
V	10	9·8	9·3	8·5	7·8	7·1	6·5

Determine the value of r.

Solution: At the point on the curve where

$$V = \frac{V_{\text{peak}}}{\sqrt{2}} = \frac{10}{\sqrt{2}} = 7\cdot1 \qquad\qquad \delta\omega = 5000$$

$$\therefore\ Q = \frac{\omega_0}{2\delta\omega} = \frac{8 \times 10^5}{2 \times 5000} = 80$$

but

$$Q = \frac{\omega L}{v}$$

$$\therefore\ v = \frac{\omega L}{Q} = \frac{8 \times 10^5 \times 1}{80 \times 1000} = 10\ \Omega$$

Example: In the circuit of Fig. 13.12(a) the source is a current generator of 1 A shunted by a resistance of 2 Ω. The coil L has

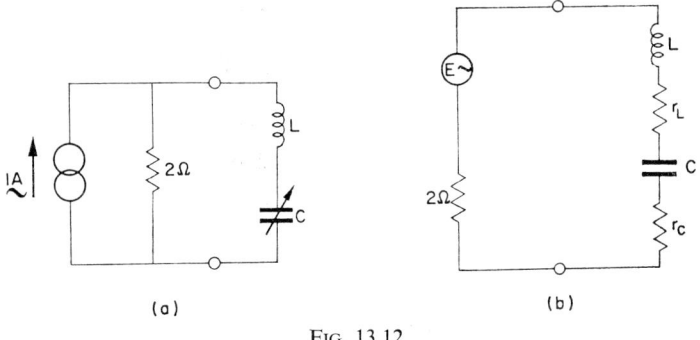

(a)   (b)

FIG. 13.12

an inductance of 1 mH and a Q-factor of 200 and C is a variable capacitor of power factor 0·002. The angular frequency of the source is 10^6 rad/sec. Determine, within 1 per cent, the voltage across C (a) when C is tuned to resonance, and (b) when, with the value of C unchanged, the frequency rises by 0·5 per cent.

Solution: Convert the current generator to a voltage generator as shown in Fig. 13.12(b) where the e.m.f. E is given by (current) × (resistance) of the current generator and the resistance is the same as for the current generator,

Include extra resistance in series to account for Q of coil and power factor of capacitor.

For coil $\qquad Q = \dfrac{X_L}{r_L}$

$$\therefore r_L = \frac{X_L}{Q} = \frac{10^6 \times 1}{200 \times 1000} = 5 \,\Omega$$

For capacitor, power factor (p.f.) $= r_C/X_C$ and $X_C = X_L$ for resonance.

$$\therefore r_C = \text{p.f.} \times X_C = 0\cdot002 \times \frac{10^6 \times 1}{1000} = 2 \,\Omega$$

$\therefore$ series resistance to be added $= 2 + 5 = 7 \,\Omega$
$\therefore$ total series resistance $= 7 + 2 = 9 \,\Omega$.
For (a) $V_C = I \times X_C$
$\qquad I = 2/9$ since reactances cancel at resonance

$$\therefore V_C = \frac{2}{9} \times \frac{10^6 \times 1}{1000} = 222 \text{ V}$$

For (b) X_L will increase by $0\cdot5$ per cent and X_C will decrease by $0\cdot5$ per cent approximately. Values will thus be as follows

$$Z = 9 + j1005 - j995 = 9 + j10 \,\Omega$$

$$|Z| = \sqrt{(9^2 + 10^2)}$$

$$\therefore |I| = \frac{2}{\sqrt{(9^2 + 10^2)}}$$

$$\therefore V_C = \frac{2}{\sqrt{(9^2 + 10^2)}} \times 995 = 148 \text{ V}$$

We have up to the present considered the efficiency of a network in passing only those frequencies required but we have not considered its effect on power transference from source to load. In many instances the precise purpose of the network is to couple the source to the load with the maximum transfer of power when the load is then said to be matched to the source.

Let us assume that we have a source of internal impedance Z_1 which is delivering power to a load of impedance Z_2 as shown in Fig. 13.13.

$$|I| = \frac{E}{\sqrt{[(R_1 + R_2)^2 + (X_1 + X_2)^2]}}$$

and if power in load is W watts

$$W = \frac{E^2 R_2}{(R_1 + R_2)^2 + (X_1 + X_2)^2}$$

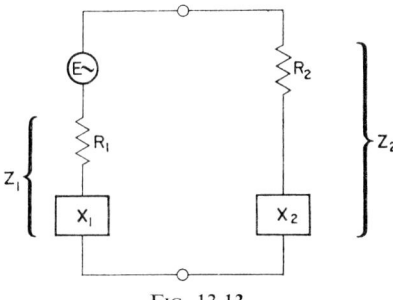

FIG. 13.13

If W is to be as large as possible it is clear that $X_1 = -X_2$ so that we can get rid of the reactive term. This gives:

$$W = \frac{E^2 R_2}{(R_1 + R_2)^2}$$

$$\frac{dW}{dR_2} = \frac{E^2(R_1 - R_2)}{(R_1 + R_2)^3} = 0 \text{ for max. } W,$$

giving $R_2 = R_1$.

Thus what is necessary for maximum power transfer is that the load shall have resistance equal to that of the source and reactance equal in magnitude and opposite in sign to that of the source.

In the practical case, however, the impedance of the load may be very different from that of the source and it is necessary to insert a network between the two so that the effective load presented by the network and the actual load has the right value of impedance.

Example: A source of internal resistance R_1 is to be matched, at angular frequency ω, to a purely resistive load R_2 (where $R_2 > R_1$) by means of the coupling circuit shown in Fig. 13.14. Derive expressions for L and C in terms of R_1, R_2 and ω.

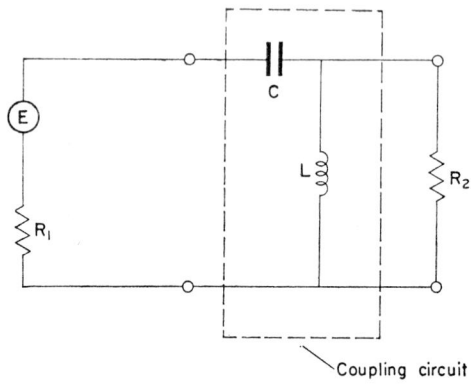

Coupling circuit

FIG. 13.14

Solution: For matching we must have:

$$R_1 = -\frac{j}{\omega C} + \frac{j\omega L R_2}{R_2 + j\omega L}$$

Equating real and imaginary parts,

$$L = CR_1R_2 \qquad \text{and} \qquad R_1\omega^2 LC = R_2\omega^2 LC - R_2$$

$$\therefore L = \frac{R_2}{\omega}\left(\frac{R_1}{R_2 - R_1}\right)$$

$$C = \frac{1}{\omega}\frac{1}{R_1(R_2 - R_1)}$$

Another type of coupling network can be designed so as to absorb a definite proportion of power otherwise dissipated in a load and at the same time to keep the loading on the source the same as before its insertion between load and source. Such a network is known as an *attenuator* and is often of the forms shown in Fig. 13.15(a) and (b). If the network is to fulfil the second condition stated above it is clear that the total input impedance of the network when the load is attached to its output terminals must be equal to the load impedance. This impedance is known as the *iterative* or *characteristic* impedance of the network.

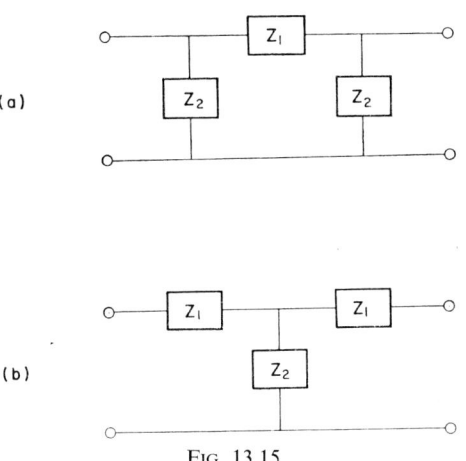

(a)

(b)

FIG. 13.15

Example: For the symmetrical Π network given in Fig. 13.15(a) derive an expression for the iterative impedance.

If Z_1 and Z_2 are pure resistors calculate their values required to give the network an iterative impedance of 100 Ω and to provide an attenuation of 20 dB when the network is inserted between a source and load each of which was a purely resistive impedance of 100 Ω.

Solution: The iterative impedance Z_0 is an impedance such that if the output terminals are connected to Z_0, the impedance looking into the input terminals is also Z_0. Hence:

$$Z_0 = \frac{Z_2\left(Z_1 + \dfrac{Z_0 Z_2}{Z_0 + Z_2}\right)}{Z_2 + Z_1 + \dfrac{Z_0 Z_2}{Z_0 + Z_2}} \quad \therefore Z_0 = \sqrt{\left(\frac{Z_1 Z_2^2}{Z_1 + 2Z_2}\right)}$$

Let input voltage $= V_1$ and output voltage $= V_2$

$\therefore$ input power $= V_1^2/Z_0$ and output power $= V_2^2/Z_0$

20 dB attenuation means that $10 \log_{10}\left(\dfrac{\text{power out}}{\text{power in}}\right) = -20$

$$\therefore \frac{\text{power out}}{\text{power in}} = \frac{1}{100}$$

$$\therefore V_1^2/Z_0 = 100V_2^2/Z_0$$

$$\therefore V_1 = 10V_2$$

From circuit
$$V_2 = \frac{\left(\dfrac{Z_0 Z_2}{Z_0 + Z_2}\right)V_1}{Z_1 + \dfrac{Z_0 Z_2}{Z_0 + Z_2}}$$

$$= \frac{Z_0 Z_2 V_1}{Z_1 Z_0 + Z_1 Z_2 + Z_0 Z_2}$$

$$\therefore 10Z_0 Z_2 = Z_1 Z_0 + Z_1 Z_2 + Z_0 Z_2$$

but
$$Z_0 = 100$$

$$\therefore 1000Z_2 = 100Z_1 + Z_1 Z_2 + 100Z_2$$

and
$$100 = \sqrt{\left(\frac{Z_1 Z_2^2}{Z_1 + 2Z_2}\right)}$$

Solution of these simultaneous equations gives:

$$Z_1 = 494{\cdot}9\ \Omega$$

$$Z_2 = 122{\cdot}2\ \Omega$$

The Π and T networks mentioned above are examples of four terminal networks with which we cannot deal generally here.

One further example is of interest as being directly applicable to coupling networks. Figure 13.16 shows a four terminal network and two external impedances Z_i and Z_o. If, when Z_o is connected alone to the output terminals, the input impedance is found to be Z_i and if, when Z_i is connected alone to the input terminals, the impedance measured at the output terminals is found to be Z_0 then Z_i and Z_0 are the *image impedances* of the network.

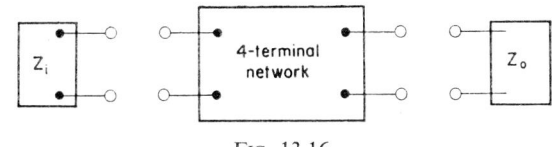

FIG. 13.16

Example: Find the image impedances for the network shown in Fig. 13.17.

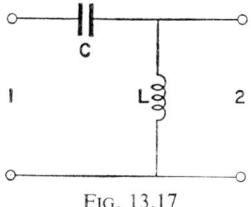

FIG. 13.17

Solution: Impedance measured at terminals 1 when Z_0 is joined to terminals 2 is:

$$\frac{1}{j\omega C} + \frac{j\omega L \times Z_0}{j\omega L + Z_0} = Z_i \qquad \text{(by definition)}$$

Impedance measured at terminals 2 when Z_i is joined to terminals 1 is:

$$\frac{[Z_i + (1/j\omega C)] \times j\omega L}{Z_i + (1/j\omega C) + j\omega L} = Z_0 \qquad \text{(by definition)}$$

Solving these two simultaneous equations

$$Z_i = \frac{1}{j\omega C}\sqrt{(1 - \omega^2 LC)} \qquad \text{and} \qquad Z_0 = \frac{j\omega L}{\sqrt{(1 - \omega^2 LC)}}$$

Typical Examination Questions

1. The circuit in Fig. 1 is used to give accentuation of bass notes in a sound reproducing system. Find values of C and R such that $|V_o/V_i|$ is four times as great at very low notes as at very high notes; and has the geometric mean of these values at middle C (256 c/s). (Thévenin's theorem simplifies the working.)

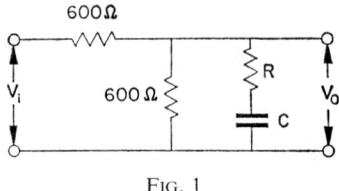

600 Ω

Fig. 1

Ans: 3·1 μF, 100 Ω

2. Derive the iterative (or characteristic) impedances of each of the networks shown in Fig. 2(a) and (b) respectively.

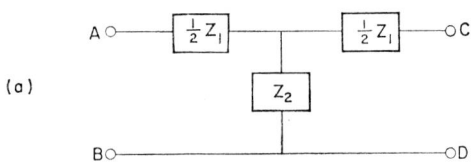

(a)

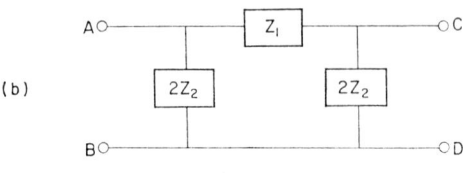

(b)

Fig. 2

Ans: $\sqrt{[Z_1Z_2+(Z_1^2/4)]}$; $\sqrt{\left[\dfrac{Z_1Z_2}{1+(Z_1/4Z_2)}\right]}$

3. For each of the circuits of Fig. 2 show that if

Z_{oc} = Impedance between A and B when C and D are open-circuited,
Z_{sc} = Impedance between A and B when C and D are short-circuited,
Z_o = Characteristic impedance,

then $Z_o = \sqrt{(Z_{oc}Z_{sc})}$,

4. The low pass filter shown in Fig. 3 is terminated in its iterative (or characteristic) impedance Z_0. Derive the value of Z_0 and hence show that no power can be fed into the termination unless the angular frequency ω lies between 0 and $2/\sqrt{(LC)}$.

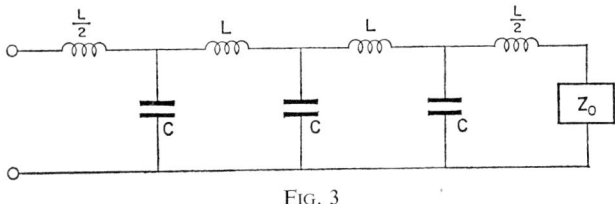

FIG. 3

5. A transformer with a primary inductance of 10 H and a secondary inductance of 1 H has a mutual inductance of 3 H. The primary circuit has a resistance of 200 Ω and is supplied from a voltage source of 100 V at 1000 c/s. The secondary side has a resistance of 20 Ω and is joined to a variable capacitor. This capacitor is adjusted to take the maximum possible current. What is this maximum current and what value of capacitor is necessary to produce it?

Ans: 1·5 A, 242 μF

6. The valve shown in the circuit of Fig. 4 drives an alternating current $i\cos\omega t$ amperes through the coil L_1. The coil is coupled by mutual inductance M to the tuned circuit consisting of L, C, and r, in series.

Find an expression for the p.d. across the condenser C.

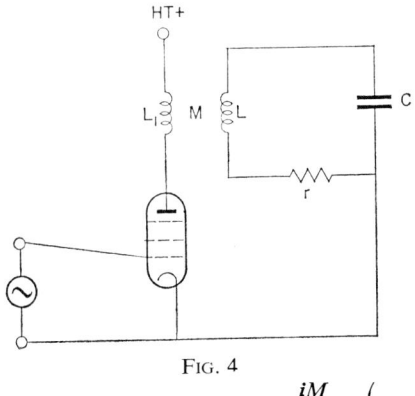

FIG. 4

$$\text{Ans:} \quad \frac{\pm\dfrac{iM}{Cr}\cos\left(\omega t - \tan^{-1}\dfrac{\omega^2 LC - 1}{\omega Cr}\right)}{\sqrt{\left[\left(\dfrac{\omega^2 LC - 1}{\omega Cr}\right)^2 + 1\right]}}$$

7. (a) A coil of 200 μH inductance has a resistance of 6·28 Ω when measured at a frequency of 1 Mc/s. What is the Q-factor of the coil at this frequency?

Ans: 200

(b) The coil is connected in series with a perfect variable condenser C and a source of e.m.f. of frequency 1 Mc/s and negligible internal resistance. C is varied until the current through the source is a maximum. By how much will the frequency of the source (at constant e.m.f.) then have to be changed to give maximum voltage across the condenser?

Ans: 6·25 c/s

(c) With the frequency kept constant at 1 Mc/s, by how much will C have to be changed to reduce the current in the circuit to one-half of its maximum value?

Ans: 1·1 pF

(d) The coil of (a) is connected in parallel with a perfect condenser C to form a parallel resonant circuit which is tuned to be purely resistive at 1 Mc/s. What is then the value of this resistance?

By how much must the frequency be varied for the complete circuit to be equivalent to a resistance in parallel with an equal reactance?

Ans: 0·251 MΩ; 2·5 kc/s

(e) If the condenser in (d) had a power factor of 0·01, what would be the effective Q of the circuit?

Ans: 67

(f) What would be the Q of the circuit if it were tuned to 1 Mc/s by stray capacity of 30 pF and power factor 0·02 in parallel with a perfect condenser of appropriate capacity?

Ans: 103

(g) What would be the Q of the circuit in (d) if a resistance of 1 MΩ were connected in parallel with the condenser?

Ans: 160

8. A generator of e.m.f. E and internal impedance $60 + j37$ Ω is connected to a load of impedance $2940 + j539$ Ω. By what factor could the transfer of power from generator to load be improved by the use of a suitable matching device?

If an ideal transformer and a single capacitor are to be used for matching, what must be the ratio of the transformer and the reactance of the capacitor?

Ans: 13·2 times; 1 : 7; -48 or -2352 Ω

9. In Fig. 5, what ratio should the ideal transformer have, in order to transfer as much power as possible from the generator E to the 15 Ω load resistance?

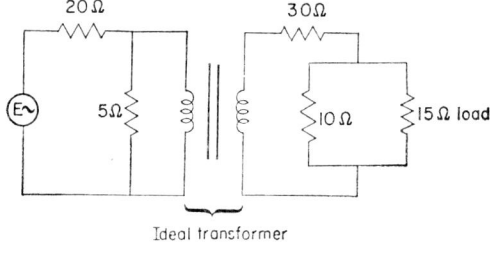

Ideal transformer

FIG. 5

Ans: 1 : 3

10. (a) An attenuator has the form of the T-network of Fig. 6. The network is terminated in a load of resistance R_o, equal to its characteristic impedance and k is the ratio of the load current to the input current. Find the values of R_1 and R_2 in terms of R_o and k.

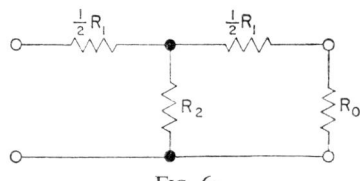

FIG. 6

Ans: $R_1 = 2R_o(1-k)/(1+k)$; $R_2 = 2kR_o/(1-k^2)$

(b) What values of R_1 and R_2 are required to produce a loss of 15 dB when $R_o = 600\ \Omega$?

Ans: $R_1 = 873 \cdot 6\ \Omega$; $R_2 = 220 \cdot 1\ \Omega$

CHAPTER 14

Polyphase Circuits

THE simplest alternating electrical supply is one in which power is drawn from two terminals where the p.d. varies sinusoidally at a constant rate. This type of supply, which is called a single-phase one, has a disadvantage, amongst others, that the power delivered fluctuates and, because the p.d. is periodically zero, drops to nothing twice per cycle.

It is, however, quite possible to combine a number of single-phase supplies so that any machine or other device run from the combination will at all instants be receiving constant power, the separate single-phase supplies being arranged to be out of phase with one another so that when one is at zero the others are not.

By far the commonest number of phases to be combined in a polyphase supply is three and we shall confine our attention in this chapter, therefore, to consideration of the solution of three-phase circuits. However, the network solutions discussed are readily applicable to any number of phases.

Let us suppose that we have three separate but identical single-phase generators (A, B, C) with their armatures mechanically coupled, but electrically independent, so that each machine has two independent terminals A_1A_0, B_1B_0 and C_1C_0. Let each generate the same e.m.f. e, of peak value E, at the frequency f. Then if e_a be the instantaneous potential of A_1 above that of A_0 (and similarly for e_b, e_c), we may write

$$e_a = E \sin \omega t$$

where $\omega = 2\pi f$, and $t = 0$ at an instant when e_a is zero and increasing. By adjusting the couplings between the armatures, rotating one relatively to the other, we could arrange that the e.m.f. of the three machines were in phase with one another. If we then unbolt the couplings and rotate armature B forwards and armature C backwards, each through one-third of a cycle, the three e.m.f. will now be:

$$e_a = E \sin \omega t,$$

$$e_b = E \sin \left(\omega t + \frac{2\pi}{3} \right) \tag{1}$$

$$e_c = E \sin \left(\omega t - \frac{2\pi}{3} \right)$$

Figure 14.1(a) shows e_a, e_b, e_c plotted against time in Cartesian coordinates; Fig. 14.1(b) shows the corresponding vector diagram with the direction of rotation marked.

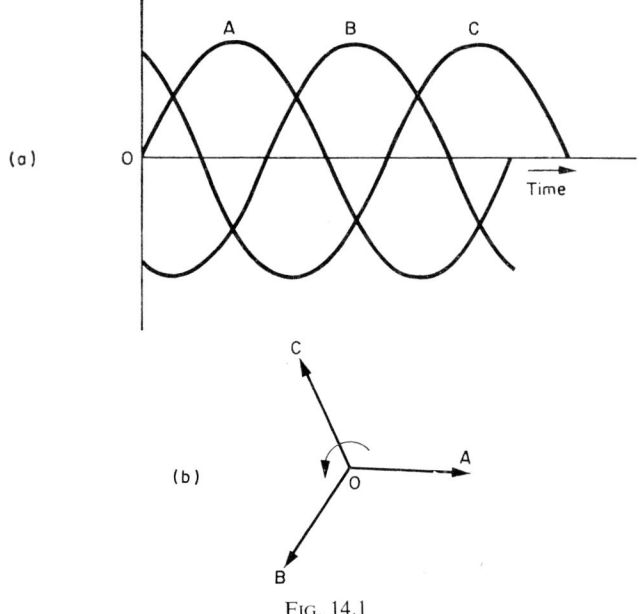

Fig. 14.1

We now have a three-phase generator although in an actual alternator we do not have three entirely separate single-phase armatures. It is more economical with regard to size and efficiency if the three independent windings are placed on the same core and rotate in the same magnetic field.

Since each winding is electrically quite separate from the others therefore let each supply equal loads, as shown in the diagram of connections, Fig. 14.2(a). We have here three entirely independent

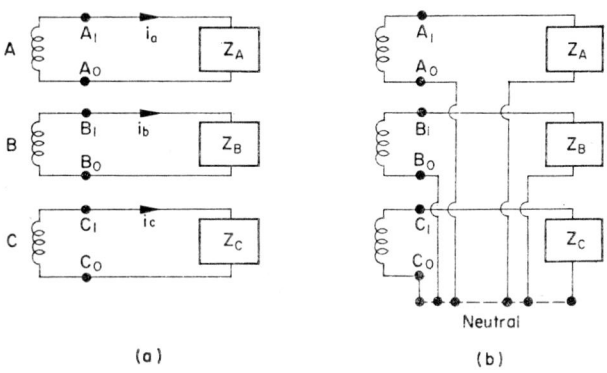

(a)   (b)

Fig. 14.2

circuits, necessitating six connecting wires. Now if we join two independent circuits at a single point, no current will flow from one circuit to the other through the joining wire. Let us then join the terminals A_0, B_0, C_0 and we can replace the three wires running from A_0, B_0, C_0 to the loads by a single wire, as shown dotted in Fig. 14.2(b). As this wire cannot be said to belong to one circuit rather than to any other, it is referred to as the "neutral"; the other wires are often referred to as "lines".

If, as we have assumed, currents are reckoned positive when flowing outwards from the neutral point of the generator, the current in the neutral wire will be the vector sum of i_a, i_b, and i_c. Since the three loads are identical, these currents will be numerically equal and will lag behind the e.m.f. e_a, e_b, e_c by equal angles as shown in the vector diagram Fig. 14.3. Adding these vectors it will be seen that the resultant is zero, so that with this arrangement of equal or "balanced" loads there is no current in the

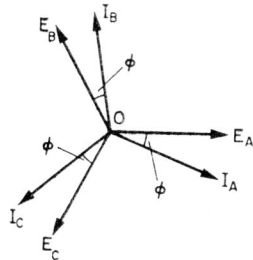

FIG. 14.3

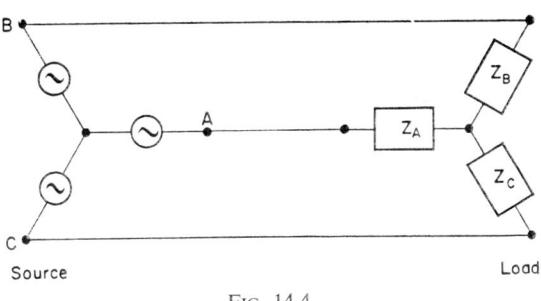

Source   Load

FIG. 14.4

neutral wire, and it may be omitted, giving the diagram of connections shown in Fig. 14.4.

Analytically the working is as follows:

$$\sin \omega t + \sin \left(\omega t + \frac{2\pi}{3}\right) + \sin \left(\omega t - \frac{2\pi}{3}\right) = 0$$

Since $\sin A + \sin B = 2 \sin \frac{1}{2}(A + B) \cos \frac{1}{2}(A - B)$

$$\therefore \sin \omega t + \sin \left(\omega t + \frac{2\pi}{3} \right) + \sin \left(\omega t - \frac{2\pi}{3} \right) = \sin \omega t + 2 \sin \omega t \cos \frac{2\pi}{3}$$

$$= \sin \omega t \left(1 + 2 \cos \frac{2\pi}{3} \right)$$

$$= 0 \quad \text{since} \quad \cos \frac{2\pi}{3} = -\tfrac{1}{2}$$

A similar proof is clearly valid if we write $(\omega t - \lambda)$ for ωt. Hence for balanced loads connected as shown, the current in the neutral is

$$i_a + i_b + i_c = I \sin (\omega t - \lambda) + I \sin \left(\omega t + \frac{2\pi}{3} - \lambda \right) + I \sin \left(\omega t - \frac{2\pi}{3} - \lambda \right)$$

$$= 0 \text{ by the above.}$$

This method of connection, usually drawn as in Fig. 14.4, is known as the *star* or *Y* method but there is an alternative method which is used as frequently as the other.

We can join two independent circuits at any point, and no current will flow in the connecting wire. Let us, in Fig. 14.2 (a), join A_0 to B_1. We have another independent circuit C, so join B_0 to C_1, and we see that the p.d. across $C_0 A_1$ is clearly at any instant the sum of the three voltages e_a, e_b, e_c which we can obtain analytically by using equations (1).

Hence, for this connection the total e.m.f. acting round the circuit is

$$e_a + e_b + e_c = E \sin \omega t + E \sin \left(\omega t + \frac{2\pi}{3} \right) + E \sin \left(\omega t - \frac{2\pi}{3} \right)$$

Then, since there is no p.d. at any time between the points C_0 and A_1, no current will flow in a wire joining them and so we can replace each pair of wires, which are now in parallel between generators and loads, by a single wire. The result is as shown in Fig. 14.5. From the shape of the diagram, drawn in its usual manner in Fig. 14.6, the system is referred to as *delta* or *mesh* connected and in this system it will be seen that, unlike the star

connection, we cannot associate any one line or supply wire with
one particular armature.

If we have a three-phase supply with no neutral wire whether
the generator is delta-connected or star-connected, the easiest
quantities to measure are the current in one line and the voltage
between any two lines. The latter, called the "line volts", V_L, will

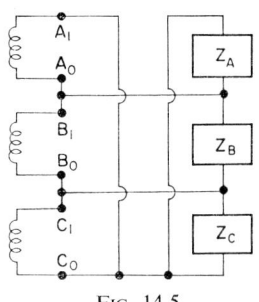

FIG. 14.5

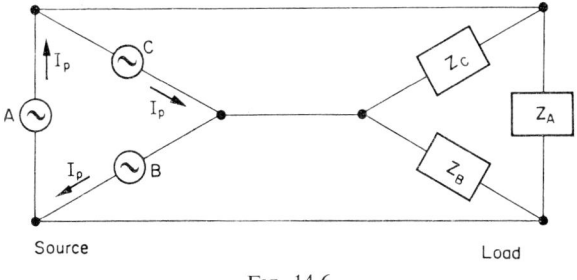

Source   Load

FIG. 14.6

be the same for any pair of lines if the drop in the generator and
lines is neglected for unbalanced loads; but in a star system this
will not be equal to the volts generated in any one winding or
"phase" of the supply, which we call the "phase volts", V_p. In a
star-connected system the line current I_L will be the same as the
current in one of the phases, I_p, but this will not be so for a delta-
connected generator. It is necessary to find the relations between
these quantities.

In the star-connected system of Fig. 14.4 the p.d. between the neutral point O and one of the lines is V_p shown in the vector diagram of Fig. 14.7 as OA,OB,OC, with the proper phase difference. Then the p.d. between A and B, or V_L, is the difference

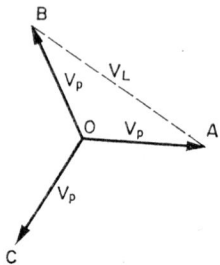

FIG. 14.7

between the p.d. of A and B referred to O, i.e. it is the vector difference between OA and OB, or AB as shown in the dotted line. Since the angle AOB is 120° or $2\pi/3$, trigonometrical calculation shows that $AB = \sqrt{3} \times OA$, or

$$V_L = \sqrt{3} V_p \qquad (2)$$

in a star-connected system.

Clearly I_p and I_L are equal, since each line carries the current from one phase. Assuming a balanced load, which means that all three phases of the load are identical, so that all armature currents I_p are equal, and lag behind the voltages V_p by the same angle λ, the power developed by each phase is

$$V_p I_p \cos \lambda$$

where $\cos \lambda$ is the *power factor*. Since there are three phases, and the load is balanced, the total power delivered will be three times this, or $W = 3 V_p I_p \cos \lambda$. If we wish to express the power in terms of line volts and currents instead of phase volts and currents, since $V_L = \sqrt{3} V_p$ and $I_L = I_p$ we have

$$W = \frac{3V_L}{\sqrt{3}} I_L \cos \lambda = \sqrt{3} V_L I_L \cos \lambda \qquad (3)$$

In a delta-connected system, Fig. 14.6, the voltage between lines V_L is equal to that of one armature or phase, so that $V_L = V_p$.

But the currents in the lines I_L are not the same as the armature currents I_p. Figure 14.8 shows the phase or armature currents

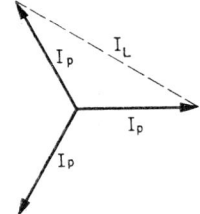

FIG. 14.8

I_p, the positive directions in the armatures being as shown in Fig. 14.6. Clearly the current in any line is the difference of two armature currents, represented by the dotted line I_L in Fig. 14.8. By trigonometry

$$I_L = \sqrt{3}I_p \qquad (4)$$

in a delta system.

As before, the power in one phase or armature is

$$V_p I_p \cos \lambda$$

and in all three the total power is three times this for balanced loads or

$$W = 3V_p I_p \cos \lambda$$

Substituting for I_p in terms of I_L gives:

$$W = 3V_L \frac{I_L}{\sqrt{3}} \cos \lambda = \sqrt{3}V_L I_L \cos \lambda$$

which is the same expression as before.

Example: A balanced three phase load of 8 kW at 0·8 power factor lagging is supplied at 440 V between lines. Find the line current and the equivalent impedance of each phase of the load if it is assumed to be in star.

Solution:
$$W = \sqrt{3}V_L I_L \cos \lambda$$

$$\therefore \quad 8000 = \sqrt{3} \times 440 \times I_L \times 0.8$$

$$\therefore \quad I_L = \frac{8000}{\sqrt{3} \times 440 \times 0.8} = 13.1 \text{ A}$$

$$V_p = \frac{V_L}{\sqrt{3}} = \frac{440}{\sqrt{3}} = 254 \text{ V}$$

$$\therefore \text{ phase impedance} = Z_p = (a+jb) = \frac{254}{13.1} = 19.4 \ \Omega$$

Now only the resistive part of the phase impedance can absorb power, so:

$$13.1^2 a = \frac{8000}{3}$$

$$\therefore \quad a = \frac{8000}{3 \times 13.1^2} = 15.6 \ \Omega$$

also
$$|Z_p| = \sqrt{(a^2+b^2)} = 19.4 \ \Omega$$

$$\therefore \quad \sqrt{(15.6^2+b^2)} = 19.4$$

$$b^2 = 19.4^2 - 15.6^2 = 133$$

$$b = 11.5 \ \Omega$$

Since the power factor is a lagging one b must be positive, giving

$$Z_p = (15.6 + j11.5) \ \Omega$$

As a check the phase angle of Z_p is

$$\tan^{-1}\left(\frac{b}{a}\right) = \tan^{-1}\frac{11.5}{15.6} = 37°$$

and a power factor of 0·8 is equivalent to an angle of

$$\cos^{-1} 0.8 = 37°.$$

Example: The same load as in the example above is supplied at 440 V between lines where each line between the generator and the load has a lagging reactance of 10 Ω and a resistance of 1 Ω. Find the generator supply voltage, kVA and power factor.

Each line is equivalent to an impedance of $(1+j10)$. This can be added to the phase impedance worked out above $(15\cdot6+j11\cdot5)$ to give a total impedance per phase of

$$(1+j10)+(15\cdot6+j11\cdot5)=(16\cdot6+j21\cdot5)$$

$$I_L = 13\cdot1\ \text{A}$$

$$\therefore\ V_p\ (\text{from generator to star point})$$

$$= 13\cdot1(16\cdot6+j21\cdot5)$$

$$= (218+j282)$$

$$\therefore |V_p| = \sqrt{(218^2+282^2)} = 356$$

$$\therefore |V_L| = 356 \times \sqrt3 = 618\ \text{V}$$

$$\text{phase angle} = \tan^{-1}\frac{282}{218} = 52°$$

$$\therefore\ \text{power factor} = \cos 52° = 0\cdot6$$

and $$\text{kilo-volt-amp} = \frac{\sqrt3(618 \times 13\cdot1)}{1000} = 14\ \text{kVA}$$

An alternative way of solving problems of this type is often known as the "reactive kVA" or "kilovar" method. Figure 14.9

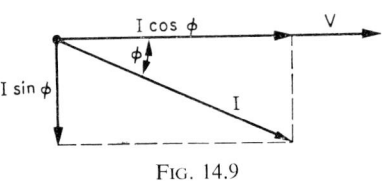

FIG. 14.9

shows, for one phase only, voltage and current vectors having phase difference ϕ. The power represented by these vectors is given by $W = VI\cos\phi$ and we define the "reactive power" as $VAR = VI\sin\phi$. We may think of this as taking the orthogonal components of I and multiplying these separately by V. We may therefore take an external view of any circuit being supplied from a sinusoidal source and obtain the two separate net sums of all the W and all the VAR components the result giving us the power consumed and the power factor.

The working of the previous example in this way is as follows: The line current I in the load is given by

$$440 \times I \times \sqrt{3} = 10^4 \qquad \therefore I = 13 \cdot 1 \text{ A.}$$

	kW	kVAR	kVA	I	V
At the load	8	6	10	13·1	440
In the line	0·516	5·16	—	13·1	—
At the supply point	8·516	11·16	14·03	13·1	618

0·516 is $3 \times 13 \cdot 1^2 \times 1 \times 10^{-3}$, 5·16 is $3 \times 13 \cdot 1^2 \times 10 \times 10^{-3}$

In line 3 add kW^2 to kVAR^2 to give kVA^2.

The volts are $440 \times \dfrac{14 \cdot 03}{10}$ since current is constant.

Selecting the figures we require for the answer we have:

$$14 \text{ kVA at 618 V and } \frac{8 \cdot 516}{14 \cdot 03} = 0 \cdot 6 \text{ p.f.}$$

When using this method with loads and supply lines having both inductive and capacitive reactance it is very important that due regard shall be given to the signs of the VAR terms.

It is clear from the above discussion that a three-phase supply may be generated by a machine connected either in star or delta and that a load may be either in star or delta. Thus a star connected machine may be connected to a star or delta load and vice versa. However, so far as analytical solutions of balanced loads are concerned the algebra may be easier if we are dealing with a connection different from that actually existing in the load. In these cases it is straightforward to convert the actual load to an equivalent one in star or delta as the case may be.

Thus if we have three equal loads $(R + jX)$ connected in delta, we can replace them by three equal loads $(r + jx)$ connected in star, so that if the respective loads were in a box with only the three terminals A, B, C available we could not tell whether the connections inside the box were Δ or Y.

To find the relation between $(R+jX)$ and $(r+jx)$ first let X and x be zero so that we have to consider R and r only. Find the resistance measured between A and B with C insulated. We have clearly in Fig. 14.10(a), R and $2R$ in parallel, which is

$$\frac{R \times 2R}{R+2R} = \frac{2R^2}{3R} = 2R/3 \ \Omega.$$

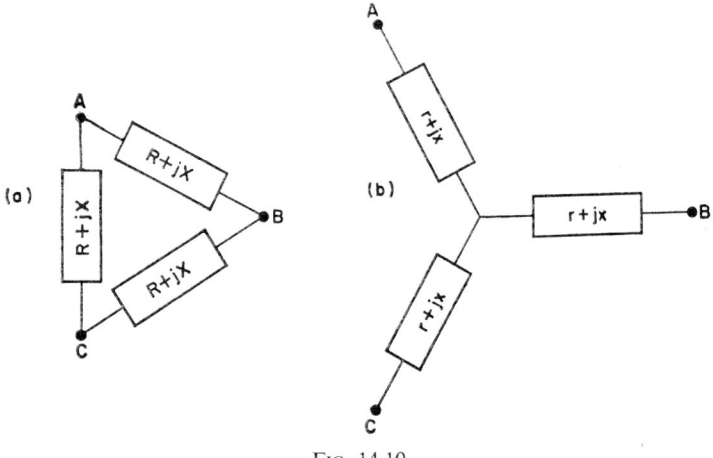

FIG. 14.10

In Fig. 14.10(b) we have r and r in series, in all $2r$. So to make the two arrangements equivalent we must have $2R/3 = 2r$, or $r = R/3$.

When reactances X and x are present we have in the j notation:

$$Z_{AB} \text{ (with } C \text{ insulated)} = \frac{2(R+jX)(R+jX)}{2(R+jX)+(R+jX)}$$

$$= \frac{2(R+jX)^2}{3(R+jX)} = 2(R+jX)/3$$

For equivalence $Z_{AB} = Z_{ab}$

$$\therefore\ 2(R+jX)/3 = (r+jx)+(r+jx)$$
$$= 2(r+jx)$$
$$\therefore\ (R+jX) = 3(r+jx)$$

$\therefore$ equating real and imaginary parts

$$R = 3r$$

and $\qquad X = 3x.$

Example: A three-phase star-connected motor delivering 800 h.p. at 92 per cent efficiency and 0·89 p.f. is supplied at 6600 V between lines and at a frequency of 50 c/s. Find the line current. In order to reduce this current to a minimum, a balanced delta-connected system of capacitors is connected in parallel with the motor. Find the value of each capacitor.

Solution: $\sqrt{3}V_L I_L \cos\phi$ = power from supply

$$\cos\phi = \text{power factor}$$

$$\therefore\ I_L = \frac{800 \times 746}{0·92 \times 0·89} \times \frac{1}{\sqrt{3} \times 6600}$$

$$= 64\text{ A}$$

$$\phi = \cos^{-1} 0·89$$

$$\therefore\ \sin\phi = \sin(\cos^{-1} 0·89) = 0·45$$

Figure 14.11 shows the vector diagram for one phase of the motor and it can be seen that the reactive component of the current is given by

$$I_r = 64\sin\phi = 64 \times 0·45$$
$$= 29·3\text{ A}.$$

If a capacitor were connected across each motor phase the current taken by each capacitor would be purely reactive and would lead the voltage by 90° as shown dotted in Fig. 14.11. If the size of capacitor is chosen so that the current through it is equal to I_r then the net current taken by the combination will be simply I_p and this will be the least possible value.

Hence
$$X_C = \frac{6600}{\sqrt{3}} \times \frac{1}{29 \cdot 3} = 130 \ \Omega$$

$$\therefore \ C = \frac{1}{\omega X_C} = \frac{1}{2\pi \times 50 \times 130} = 24 \cdot 6 \times 10^6 \ F$$

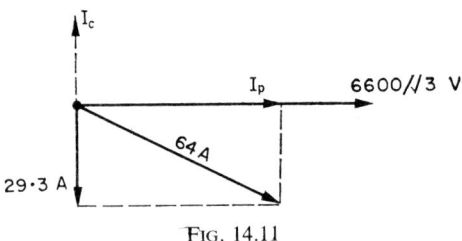

FIG. 14.11

These capacitors, being in parallel with a star load will themselves be in star. However, we are required to find the value for capacitors in delta. The equivalent delta reactance required is three times the star reactance as has been shown above.

$$\therefore \ X_C \ (\text{in delta}) = 3X_C \ (\text{in star})$$

but
$$C = \frac{1}{\omega X_C}$$

$$\therefore \ C \ (\text{in delta}) = \tfrac{1}{3}C \ (\text{in star})$$

$$= \frac{24 \cdot 6}{3} \ \mu F$$

$$= 8 \cdot 2 \ \mu F$$

Alternatively we may solve the second part of the problem as follows:

Reactive power $= VAR = \sqrt{3} \times 64 \times 0 \cdot 458 \times 6600$

$$= 336,000.$$

This must be absorbed by the three capacitors, and the reactive power in a capacitor is given by V^2/X_C.

$$\therefore 3 \times 6600^2 \times WC = 336{,}000$$

$$\therefore C = 8\cdot2\,\mu\text{F}.$$

If the phases of the load are not identical a star–delta or delta–star conversion can still be made in the manner fully discussed on pp. 61–65, but the resulting expressions are rather unwieldy and for straightforward problems involving unbalanced loads it is usually much quicker to use a graphical vector method.

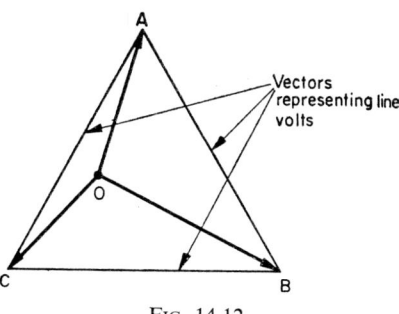

Fig. 14.12

It is quite possible for the impedances in each phase of a load to be dissimilar or unbalanced and if this is so each phase must be treated separately and, if the total power consumption is required, this will be the sum of the phase powers. The main difficulty in the analysis of unbalanced loads, particularly if they are in star, is the numerical determination of a consistent set of phase currents which will give the proper phase voltages so that vector addition of the latter gives the supply line voltages which can almost always be assumed to be balanced. Graphically the situation is easy to grasp. Figure 14.12 shows three balanced line voltages giving an equilateral triangle and a point O from which vectors have been drawn to the corners of the triangle to represent the phase voltages for a star load. It will be seen that this geometrical arrangement ensures that the phase voltages give the correct line voltages and our problem, if solved graphically, simply resolves into finding the position of O so that the phase voltages and the currents due to them in any given unbalanced load are consistent with the vector sum of the currents being zero. Hence we could proceed by trial, having chosen an arbitrary point O, to draw in the vectors representing the currents—their magnitudes and phase displacements being ascertained from the application of the appropriate voltage resulting from the choice of O—and see if these currents form a closed triangle. If not we must alter the position of O and try again.

Many numerical problems are sufficiently restricted in their data to give a simple solution without the need for trial. Thus the load in Fig. 14.13(a) can be solved by the vector diagram given in Fig. 14.13(b) if it is realized that, since the ohmic value of each phase is the same, the voltage vectors can also serve as current vectors to another scale provided that appropriate change of direction to allow for phase difference is made where necessary. No change in direction is necessary for the currents associated with the resistive phases and a rotation of 90° counter-clockwise is required for the capacitive phase. Thus the resultant of the two currents I_B and I_C must give I_A equal in length to V_A but 90° in advance of it. This resultant is the diagonal of the parallelogram and hence is bisected by CB; thus we finally require that EO shall be equal to $\frac{1}{2}AO$ and at right angles to it. Hence the locus of O is the semi-circle EOA and simple trial gives the position of point O. It should be noted that the order in which the phase voltages of the supply come to their peak positive values, that is the phase rotation of the supply, must be taken into account in drawing the diagram. If the phase rotation is given as ABC and if the current in the capacitor is taken to be 90° advanced counter-clockwise relative to the voltage across it then the diagram must be lettered as shown. If the phase rotation were reversed either the letters B and C must be reversed or else the capacitor current must be drawn 90° clockwise relative to the voltage vector.

If a short geometrical construction for a given problem is not obvious an analytical method is often preferable. The basis of this method is the designation of the line voltages in the j notation.

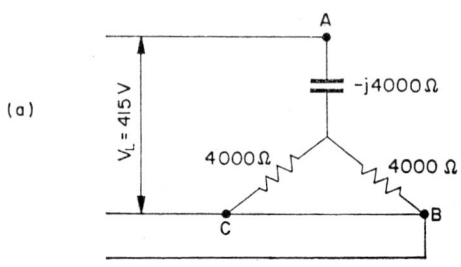

(a)

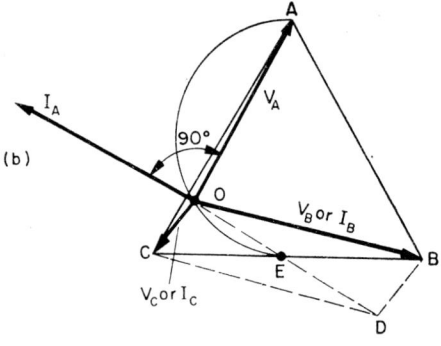

(b)

FIG. 14.13

Since relative to any one voltage the other voltages are $+120°$ or $-120°$ in phase for a balanced supply, the three supply voltages can be designated:

$$V, \quad V\left(-\tfrac{1}{2}+j\frac{\sqrt{3}}{2}\right), \quad V\left(-\tfrac{1}{2}-j\frac{\sqrt{3}}{2}\right)$$

as shown in Fig. 14.14.

We can then form the following equations:

$$415 = 4000 (I_B - I_C)$$

$$415 \left(-\tfrac{1}{2} + j\frac{\sqrt{3}}{2} \right) = 4000(I_C + jI_A)$$

$$I_A + I_B + I_C = 0$$

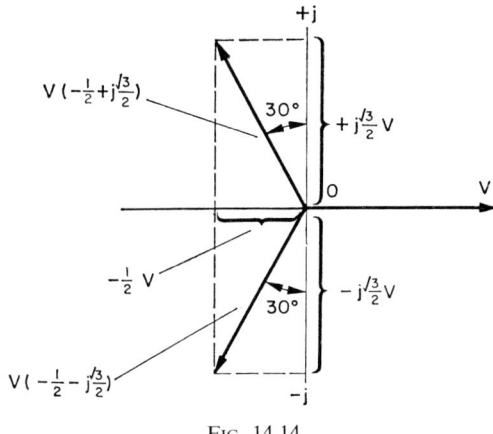

FIG. 14.14

Solving these equations we get:

$$I_A = 0 \cdot 072 - j0 \cdot 036 = 80 \text{ mA at } -26° \ 31'$$

$$I_B = 0 \cdot 016 + j0 \cdot 018 = 25 \text{ mA at } +48° \ 30'$$

$$I_C = 0 \cdot 088 + j0 \cdot 018 = 90 \text{ mA at } +168° \ 30'$$

but the algebra involved in arriving at these results is a little tedious.

Methods of measuring power in a three-phase load depend to some extent on whether the load is balanced or unbalanced and whether the neutral is available or not. If we know that the load is balanced and if we can make a connection to the neutral then one wattmeter having its current coil in series with one line and its volt coil connected between that line and the neutral will give us the power consumed in one phase of the load. Obviously the total power consumed will be three times the reading of the one wattmeter. If the neutral is not accessible or if the load is connected in delta it is possible to make an artificial star point by connecting three equal resistors in the form of a star across the supply. One end of the wattmeter volt coil is then connected to this artificial star point but it should be noted that the resistance of the resistors must be small compared with that of the watt-meter volt-coil circuit so that we may ignore the effect of the resistance of this circuit shunting one of the star resistors.

If the load is unbalanced and the neutral is available we can use three wattmeters connected to the three phases, but it is more convenient to use two wattmeters in the circuit shown in Fig. 14.15,

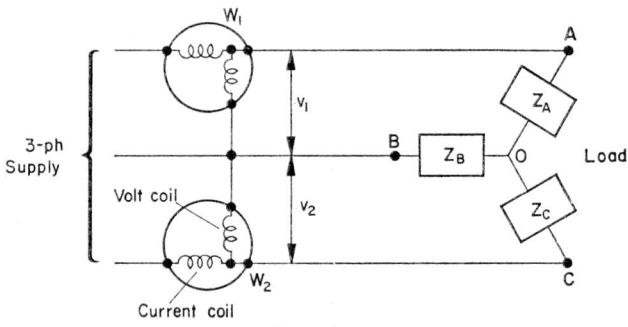

FIG. 14.15

which is the method to be used in any case if the neutral is not available. The sum of the two wattmeter readings W_1 and W_2 gives the total power consumed by the load which can be proved as follows.

Let i_a, i_b, i_c be the instantaneous values of the line currents (reckoned positive when outwards from star point of generator). Similarly, let e_a, e_b, e_c be the instantaneous p.d. across OA, OB, OC. If v_1 and v_2 are the instantaneous p.d. across the wattmeter volt coils we have:

$$v_1 = e_a - e_b, \qquad v_2 = e_c - e_b,$$

and we also have $i_a + i_b + i_c = 0$. The wattmeter powers are $w_1 = v_1 i_a$, $w_2 = v_2 i_c$ (instantaneous values). Therefore

$$w_1 + w_2 = v_1 i_a + v_2 i_c = (e_a - e_b)i_a + (e_c - e_b)i_c$$
$$= e_a i_a - e_b(i_a + i_c) + e_c i_c = \{e_a i_a + e_b i_b + e_c i_c\}.$$

But $\{\}$ is the sum of the outputs of the three phases or total power supplied. Hence at every instant the power deflecting the wattmeters is equal to the power supplied, and the same will be true when averaged over a cycle. Thus W_1, W_2 being the mean values of w_1, w_2, we have $W_1 + W_2 = $ mean rate at which power is supplied.

It will be seen that no assumptions have been made as to equality of currents or phases, so the method is valid for unbalanced loads. Indeed, it is unnecessary even to assume equality of the three voltages e_a, e_b, e_c, though a system in which these were not equal would not be employed in practice.

Example: The power supplied to a balanced three-phase load is measured by the two-wattmeter method. The wattmeter readings are $+2000$ W and -790 W. Find the power factor of the load.

Solution: Figure 14.16 shows the vector diagram for a balanced load having a power factor given by $\cos \phi$. It is clear that the wattmeters W_1 and W_2 will read $V_1 I_1 \cos \alpha$ and $V_2 I_2 \cos \beta$. From the diagram it can be seen that $\alpha = (30° - \phi)$ and $\beta = (30° + \phi)$. Thus:

$$\frac{V_2 I_2 \cos(30° + \phi)}{V_1 I_1 \cos(30° - \phi)} = \frac{-790}{2000}$$

but
$$V_1 = V_2 = V_L \text{ and } I_1 = I_2 = I_A = I_C$$

$$\therefore \frac{\cos(30° + \phi)}{\cos(30° - \phi)} = \frac{-790}{2000}$$

giving $\cos \phi = 0 \cdot 243$.

Example: A star-connected load is supplied by a three-phase supply of 400 V between lines, the phase sequence being A, B, C. The p.d. between the star point and the A, B and C lines are found to be 197 V, 273 V and 225 V respectively. It is known that the branch of the star connected to A is purely resistive. The line currents in A, B and C are 12 A, 10 A and 8 A respectively.

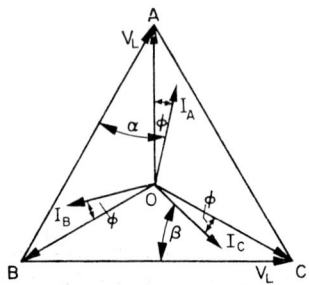

FIG. 14.16

Wattmeters are connected to read the currents in the A and B lines, the voltage coils being connected across AC and BC respectively. What are the readings of the wattmeters and the total power supplied?

Solution: From the vector diagram in Fig. 14.17, current I_A leads V_{AC} by 20°.

$$\therefore \text{ Power} = 400 \times 12 \times \cos 20° = 4 \cdot 51 \text{ kW}$$

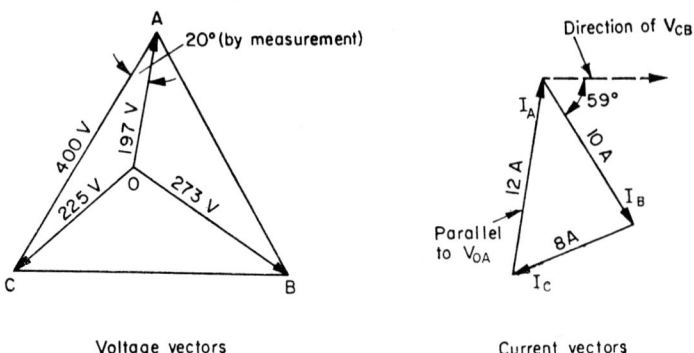

Voltage vectors   Current vectors

FIG. 14.17

Current I_B lags V_{CB} by $59°$

$\therefore$ Power $= 400 \times 10 \times \cos 59° = 2 \cdot 06$ kW

$\therefore$ wattmeters read $+4 \cdot 51$ kW and $+2 \cdot 06$ kW

$\therefore$ Total power $= 4 \cdot 51 + 2 \cdot 06 = 6 \cdot 57$ kW.

As was mentioned at the beginning of this chapter the power delivered to a balanced three-phase load is constant and this may be proved as follows:

$$e_a i_a = E \sin \omega t \times I \sin (\omega t - \lambda),$$

$$e_b i_b = E \sin \left(\omega t + \frac{2\pi}{3} \right) \times I \sin \left(\omega t + \frac{2\pi}{3} - \lambda \right),$$

$$e_c i_c = E \sin \left(\omega t - \frac{2\pi}{3} \right) \times I \sin \left(\omega t - \frac{2\pi}{3} - \lambda \right).$$

Now $\sin A \sin B = \frac{1}{2}(\cos (A - B) - \cos (A + B))$. Using this and adding:

$w = e_a i_a + e_b i_b + e_c i_c$ instantaneous watts

$$= \tfrac{1}{2} EI \left[\cos \lambda - \cos (2\omega t - \lambda) + \cos \lambda - \cos \left(2\omega t + \frac{4\pi}{3} - \lambda \right) \right.$$

$$\left. + \cos \lambda - \cos \left(2\omega t - \frac{4\pi}{3} - \lambda \right) \right]$$

$$= \tfrac{1}{2} EI \left\{ 3 \cos \lambda - \cos (2\omega t - \lambda) - \left[\cos \left(2\omega t + \frac{4\pi}{3} - \lambda \right) \right. \right.$$

$$\left. \left. + \cos \left(2\omega t - \frac{4\pi}{3} - \lambda \right) \right] \right\}$$

$$= \tfrac{1}{2} EI \left\{ 3 \cos \lambda - \cos (2\omega t - \lambda) - \left[2 \cos (2\omega t - \lambda) \cos \frac{4\pi}{3} \right] \right\}$$

$$= \tfrac{1}{2} EI \left\{ 3 \cos \lambda - \cos (2\omega t - \lambda) \left[1 + 2 \cos \frac{4\pi}{3} \right] \right\}$$

since $$\cos\frac{4\pi}{3} = -\cos\frac{\pi}{3} = -\tfrac{1}{2}$$

and therefore $$1 + 2\cos\frac{4\pi}{3} = 0.$$

$$\therefore \omega = \tfrac{1}{2}EI \times 3\cos\lambda$$

Hence if $E_{\text{r.m.s.}} = \dfrac{1}{\sqrt{2}}E$ and $I_{\text{r.m.s.}} = \dfrac{1}{\sqrt{2}}I$ are r.m.s. values per phase,

$$w = 3E_{\text{r.m.s.}}.I_{\text{r.m.s.}}\cos\lambda.$$

This expression for the instantaneous power output w is seen to be a constant, since it does not contain t.

A further advantage of a polyphase system is that it can be used to produce a magnetic field of constant strength rotating at a constant angular velocity by the application of the polyphase supply to a set of stationary coils suitably placed.

Figure 14.18 shows the disposition of the coils for a three-phase rotating field system and the constancy of the field strength and its rotation may be proved as follows:

Let $$i_a = I\sin\omega t,$$

$$i_b = I\sin\left(\omega t + \frac{2\pi}{3}\right), \qquad (5)$$

$$i_c = I\sin\left(\omega t - \frac{2\pi}{3}\right).$$

Then at any instant the magnetic field is the resultant of three components:

$$ki_a = kI\sin\omega t \qquad \text{along } OA,$$

$$ki_b = kI\sin\left(\omega t + \frac{2\pi}{3}\right) \text{ along } OB,$$

$$ki_c = kI\sin\left(\omega t - \frac{2\pi}{3}\right) \text{ along } OC.$$

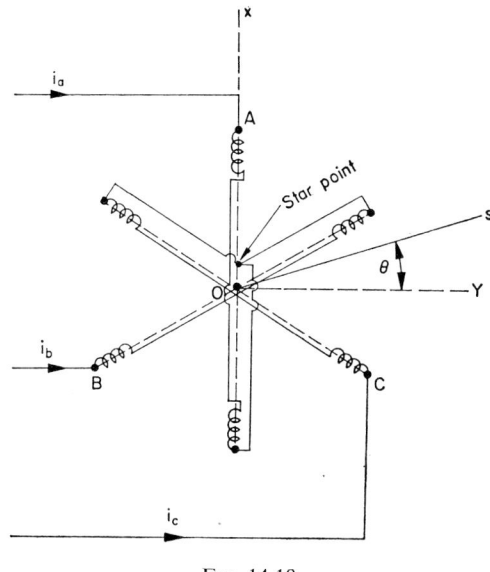

FIG. 14.18

Resolve each of these along and perpendicular to OA which is along OX and OY, and add the components in these directions giving:

Along OX:

$$kI \sin \omega t + kI \sin \left(\omega t + \frac{2\pi}{3} \right) \cos \frac{2\pi}{3} + kI \sin \left(\omega t - \frac{2\pi}{3} \right) \cos \left(- \frac{2\pi}{3} \right)$$

$$= kI \left\{ \sin \omega t - \tfrac{1}{2} \left[\sin \left(\omega t + \frac{2\pi}{3} \right) + \sin \left(\omega t - \frac{2\pi}{3} \right) \right] \right\}$$

$$\left(\text{Since} \cos \pm \frac{2\pi}{3} = - \tfrac{1}{2} \right)$$

$$= kI \left[\sin \omega t - \left(\sin \omega t \cos \frac{2\pi}{3} \right) \right]$$

$$[\text{Since } \tfrac{1}{2}(\sin A + \sin B) = \sin \tfrac{1}{2}(A + B)\cos(A - B)]$$

$$= kI \sin \omega t [1 - (-\tfrac{1}{2})]$$

$$= (3/2)kI \sin \omega t. \tag{6}$$

Along OY:

$$kI \sin\left(\omega t + \frac{2\pi}{3}\right) \sin\frac{2\pi}{3} + kI \sin\left(\omega t - \frac{2\pi}{3}\right) \sin\left(-\frac{2\pi}{3}\right)$$

$$= \frac{\sqrt{3}}{2} kI \left[\sin\left(\omega t + \frac{2\pi}{3}\right) - \sin\left(\omega t - \frac{2\pi}{3}\right)\right]$$

$$\left(\text{Since } \sin\frac{2\pi}{3} = \frac{\sqrt{3}}{2}\right)$$

$$= \sqrt{3}kI \left(\cos \omega t \sin\frac{2\pi}{3}\right)$$

$$[\text{Since } \tfrac{1}{2}(\sin A - \sin B) = \cos \tfrac{1}{2}(A + B)\sin \tfrac{1}{2}(A - B)]$$

$$= \sqrt{3}\, kI\, (\cos \omega t) \left(\frac{\sqrt{3}}{2}\right)$$

$$= (3/2)\, kI \cos \omega t \tag{7}$$

Now consider a magnetic field of constant strength $(3/2)kI$ acting along the direction OS, Fig. 14.18, making an angle θ with OY, rotating with angular velocity ω in a clockwise direction so that $\theta = \omega t$; the components along OX and OY will be exactly the components given by equations (6) and (7). Thus a constant magnetic field rotating with angular velocity $\omega = 2\pi f$ making f rev/sec is equivalent to our polyphase field. Note that if we have $2P$ poles per phase instead of only 2, the speed of rotation will of course be f/P rev/sec instead of f.

Typical Examination Questions

1. A three-phase four-wire system, with phase voltage 200 V, supplies a star-connected load consisting respectively of a resistance of 20 Ω, an inductor of 25 Ω and a capacitor of 30 Ω. Determine the current in the neutral line.

Ans: $(2 \cdot 7 - j0 \cdot 67)$ A

2. A three-phase system supplies a mesh-connected load Z_1 $(R = 125 \ \Omega)$, Z_2 $(R = 155 \ \Omega, L = 0 \cdot 35$ H in series) and Z_3 $(C = 24 \cdot 6 \ \mu F)$ at 550 V, 25 c/s. Determine the three line currents. The phase sequence is 1–2–3.

Ans: $(6 \cdot 24 + j1 \cdot 06)$ A, $(-6 \cdot 95 - j21 \cdot 17)$ A, $(0 \cdot 71 + j1 \cdot 11)$ A

3. A 440 V, three-phase three-wire system supplies a star-connected load with the following branch impedances: $Z_A = 100 \ \Omega$, $Z_B = j5 \ \Omega$, $Z_C = -j5 \ \Omega$. Calculate the voltage drop across each branch and the potential of the star point with respect to the neutral of the system. The phase sequence is A–B–C.

Ans: -8550 V, 8804 V, $8426 \lfloor -1° \ 30'$, $8426 \lfloor 1° \ 30'$

4. Impedances $Z_A = (100 + j0) \ \Omega$, $Z_B = [0 - j(100/\sqrt{3})] \ \Omega$ and $Z_C = (r + jx) \ \Omega$ are connected in star across a balanced three-phase supply (phase sequence A–B–C). Find the value of Z_C if the voltages across the impedances are balanced.

Ans: $(28 \cdot 6 - j24 \cdot 8) \ \Omega$

5. Determine an expression for obtaining power factor, $\cos \phi$, by the ratio of the readings of two wattmeters connected as shown in Fig. 1 in a balanced three-phase star-connected system. Both current coils and one voltage coil are connected respectively in and across one phase, the second voltage coil being connected across the other two lines.

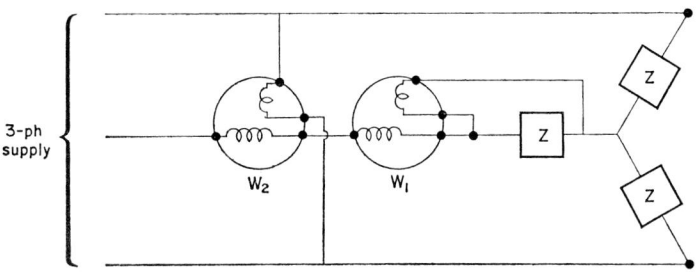

Fig. 1

Ans: $\dfrac{1}{\sqrt{3} \tan \phi} = \dfrac{w_1}{w_2}$ hence $\cos \phi$

6. A three-phase, four-wire system, with a phase voltage of 230 V, supplies a star-connected load of $Z_1 = (25\cdot0 + j21\cdot2)\ \Omega$, $Z_2 = (15\cdot5 + j17\cdot0)\ \Omega$ and $Z_3 = (9\cdot5 - j15\cdot75)\ \Omega$. Impedances $Z_{31} = (45 + j66)\ \Omega$, $Z_{23} = (100 + j87\cdot5)\ \Omega$, are also connected across the lines 1–3 and 2–3 respectively. Calculate the line and neutral currents.

Ans: $(5\cdot73 - j9\cdot51)$ A, $(-11\cdot74 - j4\cdot4)$ A, $(-10\cdot9 + j7\cdot46)$ A, $I_n = (16\cdot91 + j6\cdot44)$ A

7. A three-phase, 50 c/s, distribution line, approximately 45 miles long, delivers 1500 kW at 66 kV and 0·85 power factor lagging. Each conductor has a resistance of 7·83 Ω and the inductance and capacitance of the line are 86 mH and 0·706 μF respectively (line to neutral). Find, concentrating all line capacitance at the load end, the line voltage and current at the sending end.

Ans: 150 A, 71·5 V

8. Three sinusoidal generators 1, 2, 3 have e.m.f. represented by 200 e^{j100t}, 2000 $e^{j[100t+(2\pi/3)]}$, 2000 $e^{j[100t-(2\pi/3)]}$. They supply impedances, as shown in Fig. 2. Find the potential w.r.t. earth of the star point P, and thence the current I.

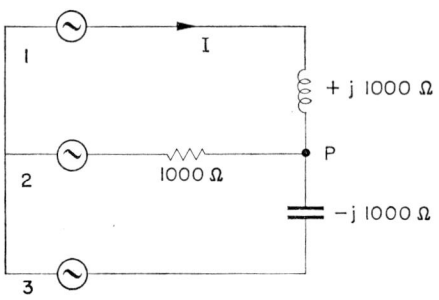

FIG. 2

Ans: V_P is represented by $2000(1 + \sqrt{3})\ e^{j100t}$, I by $[\sqrt{3} + (3 + 2\sqrt{3})j]\ e^{j100t}$.

CHAPTER 15

Simple Transients

Up to now we have been considering only voltages and currents which have been varying cyclically with time and which have been of the form $v = V \sin \omega t$ or $i = I \sin \omega t$. However before any circuit can get into its steady state of operation there must in general be a period during which the voltages and currents are changing in a non-cyclic manner and their values during this period are known as transient values. It will be clear, for instance, that if we have a capacitor connected through a resistor to a source of steady p.d. by means of a switch as shown in Fig. 15.1(a),

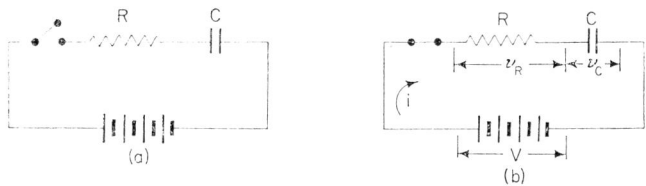

Fig. 15.1

the p.d. across the capacitor cannot instantly rise to that of the source when the switch is closed as shown in (b). There must be a period during which the charge flows into the capacitor and during which the p.d. across the capacitor builds up. Similarly, if an inductor is connected through a resistor to a source as shown in Fig. 15.2(a) and (b) a certain time is required for the establishment of the magnetic field and consequent storage of energy.

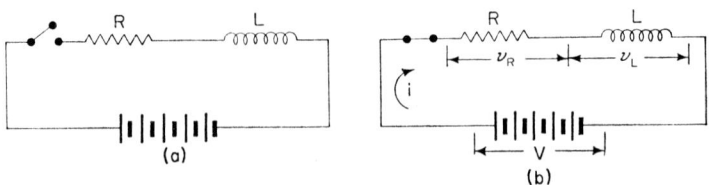

FIG. 15.2

FIG. 15.2

The relationship between the current and voltage and time during the transient period is most conveniently expressed by means of differential equations, for it will be remembered from earlier chapters that for a capacitor:

$$It = Q = VC$$

$$\therefore \quad i\,dt = dQ = C\,dv$$

$$\therefore \quad i = C\frac{dv}{dt}$$

$$\therefore \quad \text{also} \quad dv = \frac{1}{C}\,i.dt$$

$$\therefore \quad v = \frac{1}{C}\int i.dt$$

and similarly for an inductor:

$$\text{Back e.m.f.} = e = L\frac{di}{dt}$$

$$\therefore \quad e.dt = L.di$$

$$\therefore \quad i = \frac{1}{L}\int e.dt$$

We must also remember that at all times the instantaneous p.d. across a resistor is given by:

$$e = iR$$

and the instantaneous current through it by:

$$i = \frac{e}{R}$$

The above expressions will enable us to write down equations for the instantaneous currents and voltages in a large variety of circuits made up of inductors, capacitors and resistors, and the solutions of these differential equations will enable us to obtain expressions for both the transient and steady state of any required parameter. The equations resulting from this method of analysis may be easy or difficult to solve, but once the equations have been formed correctly the successful solution of the circuit has been implicitly obtained.

In what follows we shall first consider very simply examples of transients where a source of constant p.d. is suddenly connected to the circuit. More complicated circuits including those with sinusoidal sources of p.d. are discussed in detail in the next chapter.

Referring again to Fig. 15.1(b) let the instantaneous current in the circuit at any time t after closing the switch be i.

Then the instantaneous p.d. across the resistor is given by:

$$v_R = iR$$

and the instantaneous p.d. across the capacitor is given by:

$$v_C = \frac{1}{C} \int i \, dt$$

$$\therefore \quad V = iR + \frac{1}{C} \int i \, dt$$

$$\therefore \quad \frac{dV}{dt} = R \frac{di}{dt} + \frac{i}{C}$$

$$\text{but} \quad V = \text{const} \quad \therefore \quad \frac{dV}{dt} = 0$$

$$\therefore \quad 0 = R\frac{di}{dt} + \frac{i}{C}$$

$$\therefore \quad dt = -CR\frac{di}{i}$$

$$\therefore \quad t = -CR\int \frac{di}{i} = -CR\log_e i + \text{const.}$$

We must now evaluate the constant and this is done by putting a set of known simultaneous values into the equation, the set of values being called boundary conditions. In this case when $t = 0$ there is no charge on the capacitor hence:

$$i = \frac{V}{R} \quad \text{when} \quad t = 0$$

and substituting these values in the equation we get:

$$\text{const.} = CR\log_e \frac{V}{R}$$

$$\therefore \quad t = -CR\log_e i + CR\log_e \frac{V}{R}$$

$$\therefore \quad -\frac{t}{CR} = \log_e i - \log_e \frac{V}{R}$$

$$\therefore \quad e^{-t/CR} = \frac{i}{V/R}$$

$$\therefore \quad i = \frac{V}{R}e^{-t/CR}$$

Now the maximum value of i is its initial value which must be V/R because, with no charge across the capacitor, there can be no p.d. across it and so all the applied p.d. must be across the resistor. Calling this maximum value I we get:

$$i = I\,e^{-t/CR}$$

as an alternative expression giving the subsequent current in terms
of the initial one.

The curve relating i and t plotted from the above expression is
shown in Fig. 15.3 and it will be seen that a tangent to the curve

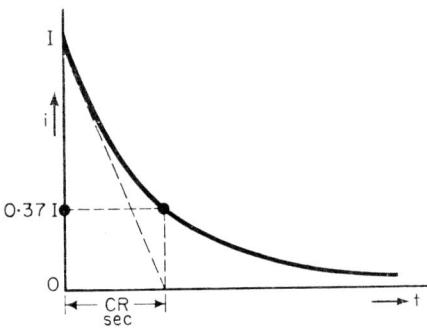

Fig. 15.3

at $t = 0$ cuts the time scale CR seconds from $t = 0$. This is easily
verified from the equation:

$$0 = R\frac{\mathrm{d}i}{\mathrm{d}t} + \frac{i}{C}$$

$$\therefore \quad \frac{\mathrm{d}i}{\mathrm{d}t} = -\frac{i}{CR} \quad \text{and} \quad \text{at } t = 0, \qquad i = I$$

$$\therefore \quad \text{slope} = -\frac{I}{CR}$$

which is the slope of the tangent in Fig. 15.3. This means that the
current would have died away to zero in time CR seconds if its
rate of fall had been maintained at its initial rate.

It is interesting to determine the actual value of the current at time CR seconds from $t = 0$ and this can be done by substituting CR for t in the equation for i. This gives:

$$i = I \, e^{-CR/CR} = I \, e^{-1}$$

$$= 0 \cdot 37I$$

Thus the current has fallen to about one third of its initial value in time CR seconds.

CR is known as the *time-constant* of the circuit and is very useful in giving a quickly calculated indication of the rate at which changes will take place in a circuit.

In the above analysis we have determined i but v_C or v_R can also be found readily. It will be seen that:

$$v_R = iR$$

$$\therefore \quad v_C = V - v_R = V - iR$$

$$\therefore \quad v_C = V - I \, e^{-t/CR} . R \quad \text{and} \quad I = \frac{V}{R}$$

$$v_C = V - \frac{V}{R} e^{-t/CR} . R$$

$$= V(1 - e^{-t/CR})$$

For v_R we have

$$v_R = iR = R . I \, e^{-t/CR}$$

Referring now to the circuit shown in Fig. 15.2(b) we have the case of an inductor in series with a resistor. The differential equation may be written down directly as:

$$V = iR + L \frac{di}{dt}$$

$$\therefore \quad \frac{V}{R} = i + \frac{L}{R} \frac{di}{dt}$$

and $V/R = I$ the final current in the circuit when all change in current has ceased so that the equation becomes:

$$I = i + \frac{L}{R}\frac{di}{dt}$$

$$\therefore \quad (I - i) = \frac{L}{R}\frac{di}{dt}$$

$$\therefore \quad \frac{R}{L}dt = \frac{1}{(I - i)}di$$

hence integrating we have

$$\frac{R}{L}t = -\log_e(I - i) + \text{const.}$$

and to obtain the constant we use initial condition $i = 0$ when $t = 0$

$$\therefore \quad 0 = -\log_e I + \text{const}$$

$$\therefore \quad \text{const.} = \log_e I$$

$$\therefore \quad \frac{R}{L}t = -\log_e(I - i) + \log_e I$$

$$\therefore \quad -\frac{R}{L}t = \log_e\left(\frac{I - i}{I}\right)$$

$$\therefore \quad e^{-(R/L)t} = \frac{I - i}{I}$$

$$\therefore \quad I\,e^{-(R/L)t} = I - i$$

$$\therefore \quad i = I(1 - e^{-(R/L)t})$$

and for v_R we have

$$v_R = V - iR$$
$$= V - IR(1 - e^{-(R/L)t})$$

but $\quad IR = V \quad \therefore \quad v_R = V - V + V\,e^{-(R/L)t}$
$$= V\,e^{-(R/L)t}$$

and $\qquad\qquad v_L = V - v_R$

The time-constant in this case is L/R and it can be seen from the curve for current against time shown in Fig. 15.4 that if the initial rate of rise of current were maintained the maximum value $I = V/R$ would be attained in a time L/R seconds from $t = 0$. Actually the current rises to 0·63, or roughly two thirds, of its maximum value in time L/R seconds and for an inductive circuit evaluation of L/R is just as useful as evaluation of CR for a capacitive one.

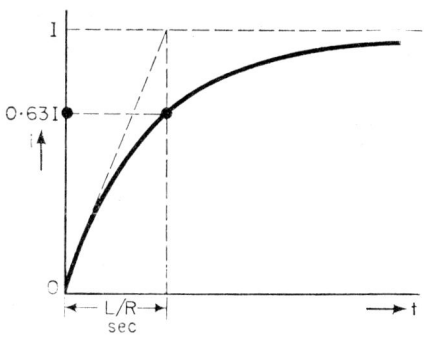

Fig. 15.4

More complicated circuits are analysed in exactly the same way as the simple circuits described above but it will be clear that the differential equations tend to become more complex and the solutions more varied and possibly more difficult to obtain.

As an example of a more complicated differential equation we shall consider the circuit shown in Fig. 15.5 where R, C and L are all connected in series. In forming the equation for this circuit it is convenient to let q be the instantaneous charge on the capacitor and to remember that, since $i.dt = dq$, then $i = dq/dt$.

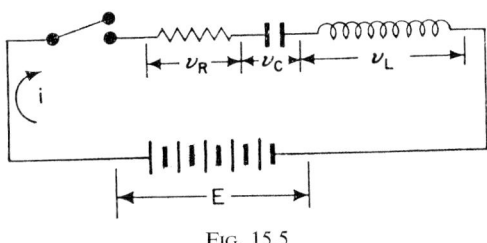

FIG. 15.5

Therefore summing the instantaneous potentials across each component and equating the sum to the applied p.d. we get:

$$L\frac{di}{dt} + Ri + \frac{q}{C} = E$$

$$\therefore \quad L\frac{d^2q}{dt^2} + R\frac{dq}{dt} + \frac{q}{C} = E$$

and we now want the complementary function and particular integral for this equation. Using the normal methods detailed in any textbook of mathematics we find that there are three possible solutions for the complementary function:

(a) $R^2 > \dfrac{4L}{C}$ c.f. $= A\,e^{-\alpha t} + B\,e^{-\beta t}$

where α and β are roots of the auxiliary equation

$$Lm^2 + Rm + \frac{1}{C} = 0$$

(b) $R^2 < \dfrac{4L}{C}$ c.f. $= e^{-\mu t}(A\cos \omega t + B\sin \omega t)$

where $\mu = -\dfrac{R}{2L}, \quad \omega^2 = \sqrt{\left(\dfrac{1}{LC} - \dfrac{R^2}{4L^2}\right)}$

(c) $R^2 = \dfrac{4L}{C}$ c.f. $= A\,e^{-\mu t} + Bt\,e^{-\mu t}$

and, if there is no charge on the capacitor at the time when the switch is closed, $t = 0$, the boundary conditions are:

$$t = 0, \qquad q = 0, \qquad \frac{dq}{dt} = 0.$$

To obtain the particular integral assume that, because E is constant, q is constant and equals D. Then

$$\frac{d^2q}{dt^2} = 0, \qquad \frac{dq}{dt} = 0$$

and original equation reduces to:

$$0 + 0 + \frac{D}{C} = E$$

$$\therefore \quad D = CE$$

Thus the complete solution is given by:

$$q = CE + \text{complementary function (a), (b) or (c)}$$

If we take case (b) as an example we get:

$$q = CE + e^{-\mu t}(A \cos \omega t + B \sin \omega t)$$

and we now have to evaluate A and B from the boundary conditions $t = 0$, $q = 0$, $dq/dt = 0$.

Substituting directly we get:

$$0 = CE + 1(A + 0)$$

$$\therefore \quad A = -CE$$

Differentiating we get:

$$\frac{dq}{dt} = 0 + (e^{-\mu t}. -\omega A \sin \omega t) + (-\mu e^{-\mu t}. A \cos \omega t)$$

$$+ (e^{-\mu t}.\omega B \cos \omega t) + (-\mu e^{-\mu t}B \sin \omega t)$$

$$\therefore \quad 0 = 0 + 0 + (-\mu A) + (\omega B) + 0$$

$$0 = (-\mu . -CE) + \omega B$$

$$\therefore \quad B = -\frac{\mu CE}{\omega}$$

and thus for the final solution we have

$$q = CE + e^{-\mu t}\left(-CE \cos \omega t - \frac{\mu CE}{\omega} \sin \omega t\right)$$

$$\therefore \quad \frac{q}{C} = v_C = E\left[1 - e^{-\mu t}\left(\cos \omega t + \frac{\mu}{\omega} \sin \omega t\right)\right]$$

It is clear from this that as t increases the equation finally reduces to $v_C = E$ which is what would be expected from inspection of the circuit since, when the current has died away, there can be no p.d. across L or R. The variation in v_C expressed in the above equation is shown graphically in Fig. 15.6 where it can be seen that the initial oscillations die away leaving the capacitor charged to a steady potential E volts.

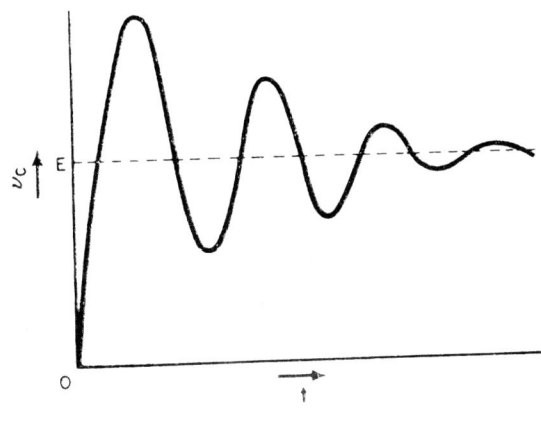

FIG. 15.6

Typical Examination Questions

1. A resistance of 1000 Ω in series with a 1500 μF condenser is suddenly connected across a d.c. supply of 300 V. Establish an expression for the voltage across the condenser after a time t seconds.

If the condenser has a 500 Ω resistance connected across its terminals before the supply is connected, derive the new expression for the voltage. What is the time-constant in the second case?

Ans: $300(1 - e^{-0.67t})$; $100(1 - e^{-2t})$; $\frac{1}{2}$ sec.

2. If a 1000 μF condenser in series with a 500 Ω resistance is placed across the 1500 μF condenser of question 1, establish expressions for the p.d. across the second condenser,

(a) if it is initially uncharged, and the supply is switched on,

(b) if, after reaching its final steady potential, the supply is removed and the 1000 Ω resistance is connected across the larger condenser.

3. A 1 μF condenser, charged to a voltage of 50 V, discharges through a resistance of 1 MΩ. How long does it take to discharge to (1) 25 V; (2) 2·5 V?

Ans: 0·69 sec., 3 sec.

4. A pair of condensers, capacity C each, are charged to voltages V_1 and V_2. They are then connected together through a resistance R, at the instant $t = 0$ as in Fig. 1. Find an expression for the current in R. From this show that the whole of the change in stored energy in the condensers before and after connexion appears as heat in R.

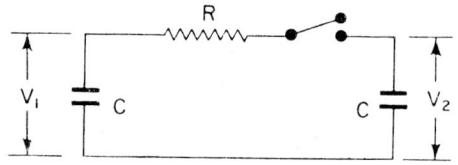

FIG. 1

Ans: $i = \dfrac{V_1 - V_2}{R} e^{-t/CR}$

5. Put down the differential equation for i the circuit in Fig. 2 after the switch has closed. What is the time constant?

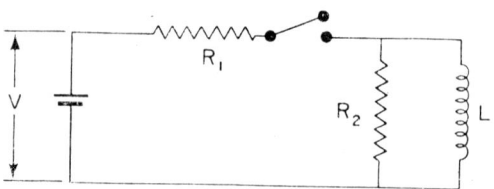

FIG. 2

Ans: $\dfrac{di}{dt} + \dfrac{R_1 R_2}{(R_1 + R_2)L} i = \dfrac{R_2}{(R_1 + R_2)} \dfrac{V}{L}, \quad \dfrac{L(R_1 + R_2)}{R_1 R_2}$

CHAPTER 16

Non-sinusoidal Excitation

VOLTAGES and currents having waveforms of sinusoidal form are common, and as has been shown in the previous chapters, lend themselves very readily to vector representation graphically or analytically. However, many other waveforms are found for these parameters, whether unintentionally due to inevitable imperfections in generating apparatus or intentionally as an essential part of the working of the apparatus. We must therefore be able to calculate the response of a network to waveforms of the types shown in Fig. 16.1 which are (a) square wave, (b) triangular wave, (c) step-function, (d) square pulse, and there can be, of course, an infinite number of actual forms. We can distinguish between the (a) or (b) type and (c) or (d) type by noticing that it is implied that (a) and (b) are repetitive indefinitely whereas (c) and (d) are "once for all" changes.

It is quite possible to adopt a general differential calculus method of expressing the network response to any type of waveform but it will usually be found easier to deal with the periodic forms by one method and the transient forms by another.

Thus when dealing with periodic forms it may only be necessary to split the waveform up into its sinusoidal components by the use of Fourier's theorem and to apply each of these components by itself to the network using any of the methods previously discussed and then to sum the results to get the final response of the network. In many instances the driving voltage or current waveform may be known to consist of a fundamental and one or two harmonics thus making the analysis simpler still.

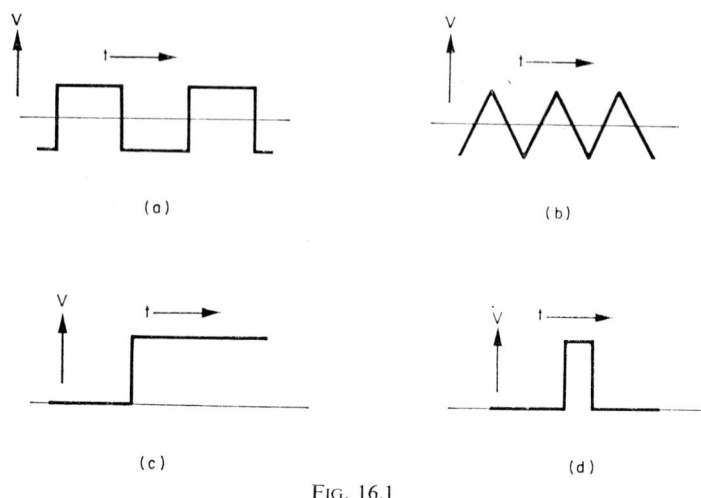

(a)

(b)

(c)

(d)

FIG. 16.1

Example: For the circuit given in Fig. 16.2 the applied e.m.f. is given by:

$$E = 10 + 12 \sin 2\pi ft + 9 \sin [4\pi ft + (\pi/4)]$$

and L is a pure inductance whose reactance at frequency f is 2 Ω and C is a pure capacitance whose reactance at frequency f is 6 Ω. The resistance of R is 5 Ω.

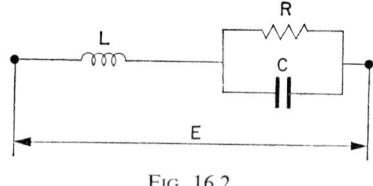

FIG. 16.2

Find the steady-state value of the current, the r.m.s. values of the two components of alternating current and the r.m.s. value of the total current taken from the supply.

Solution: The reactive components have no effect on the steady component of current due to the steady component of p.d.

$$\therefore \qquad I_{DC} = \frac{E_{DC}}{R} = \frac{10}{5} = 2 \text{ A}$$

At frequency f,

$$\omega L = 2 \, \Omega, \qquad \frac{1}{\omega C} = 6 \, \Omega$$

$$\therefore \quad I_{f(\text{peak})} = \frac{12}{j2 - (5 \times j6)/(5 - j6)} = \frac{3(5 - j6)}{3 - j5}$$

$$\therefore \quad I_{f(\text{peak})} = \frac{3\sqrt{(25 + 36)}}{\sqrt{(9 + 25)}} = 4 \cdot 02 \text{ A}$$

$$\therefore \quad I_{f(\text{r.m.s.})} = \frac{4 \cdot 02}{\sqrt{2}} = 2 \cdot 84 \text{ A}$$

At frequency $2f$,

$$\omega L = 4 \, \Omega, \qquad \frac{1}{\omega C} = 3 \, \Omega$$

$$\therefore \quad I_{2f(\text{peak})} = \frac{9}{j4 - (5 \times j3)/(5 - j3)} = \frac{9(5 - j3)}{12 + j5}$$

$$\therefore \quad I_{2f(\text{peak})} = \frac{9\sqrt{(25 + 9)}}{\sqrt{(144 + 25)}} = 4 \cdot 04 \text{ A}$$

$$\therefore \quad I_{2f(\text{r.m.s.})} = 2 \cdot 85 \text{ A}$$

Total $\qquad I_{(\text{r.m.s.})} = \sqrt{(2^2 + 2 \cdot 84^2 + 2 \cdot 85^2)}$

$$= 4 \cdot 5 \text{ A}$$

It should be noted that the phase displacement of $\pi/4$ between the fundamental and harmonic components makes no difference to the r.m.s. value. It does, however, make a difference to the combined waveform as is shown in Fig. 16.3 where the effect of adding the same two components with different phase displacements is clearly seen.

It should also be noted that the r.m.s. value of the combination of several sinusoidal waves of different frequency is the root of the sum of the mean squares and not the sum of the root mean squares.

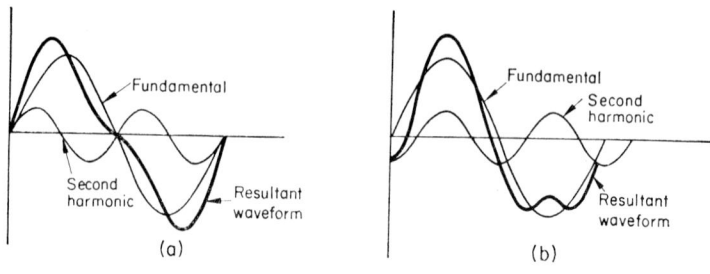

FIG. 16.3

Example: The circuit shown in Fig. 16.4(a) has a Q of 10 and carries a current of the waveform shown in Fig. 16.4(b). The peak amplitude of each pulse is 1 A and the waveform is given by:

$$I = \cos \frac{3\omega t}{2} \quad \text{for} \quad -\frac{\pi}{3} \leqslant \omega t \leqslant \frac{\pi}{3}, \text{ etc.}$$

$$I = 0 \qquad \text{for} \quad \frac{\pi}{3} \leqslant \omega t \leqslant \frac{5\pi}{3}, \text{ etc.}$$

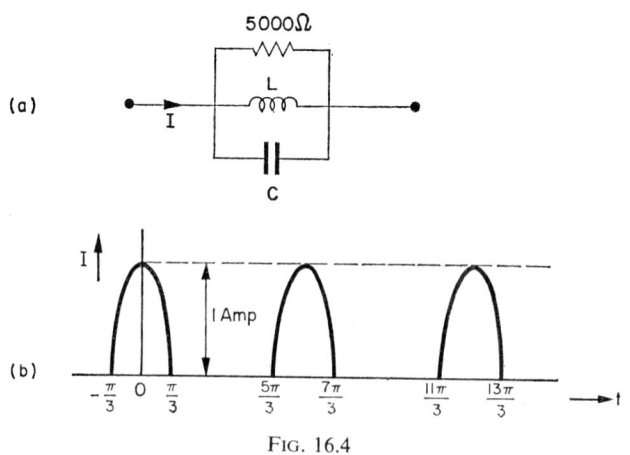

FIG. 16.4

R has a resistance of 5000 Ω and the circuit is tuned to the fundamental frequency of the current. Find the power dissipated in R by the fundamental component of the current and the approximate power dissipated by the second harmonic of the current.

Solution: By Fourier's theorem the current is of the form:

$$I = a_0 + a_1 \cos \omega t + a_2 \cos 2\omega t \ldots$$

giving $\qquad a_1 = \dfrac{6}{5\pi} \qquad$ and $\qquad a_2 = \dfrac{6}{7\pi}$

At fundamental frequency, to which the circuit is resonant, all the external current I flows through R.

$$\therefore \ W_f = \left(\frac{I_{f(\text{peak})}}{\sqrt{2}}\right)^2 \times R$$

$$= \tfrac{1}{2}\left(\frac{6}{5\pi}\right)^2 \times 5000 = 365 \text{ W}$$

Since the Q-factor is 10, at the fundamental frequency

$$X_L = X_C = \frac{R}{10} = 500 \ \Omega$$

$\therefore$ at the second harmonic frequency

$$X_L = 2 \times j500 = j1000 \ \Omega$$

$$X_C = \tfrac{1}{2} \times (-j500) = -j250 \ \Omega$$

$\therefore$ L and C in parallel have reactance

$$\frac{j1000(-j250)}{j1000 - j250} = -333 \ \Omega$$

This is much less than R, so nearly all the second harmonic current will flow through L and C and the peak voltage across the circuit will be

$$333 \times \frac{6}{7\pi} = 91 \text{ V}$$

The second harmonic power is therefore given by

$$W_{2f} = \left(\frac{V_{2f\text{(peak)}}}{\sqrt{2}}\right)^2 \times \frac{1}{R}$$

$$= \tfrac{1}{2}\left(\frac{91^2}{5000}\right) = 0\cdot8 \text{ W}$$

Waveforms of the type discussed above may be produced in a variety of ways and they are often the result of non-linear circuit elements. A non-linear circuit element may be defined as a device whose impedance varies with the current flowing or the voltage applied. Thus we might have for an iron-cored inductor:

$$V = L\frac{di}{dt} \qquad \text{where} \qquad L = f(i)$$

the effect being due to saturation of the core causing a reduction of the value of L for the higher values of current.

The basic non-linear element may be considered to be the perfect diode or rectifier which will pass current with zero voltage drop in one direction and behave as a perfect insulator in the other direction.

The addition of linear elements to non-linear ones can give us the waveform we require if the non-linear characteristic is unsuitable. Figure 16.5 gives a few very simple circuits using non-linear elements and shows the resulting waveforms.

We shall next consider the effect on a network of a non-repetitive change in the applied p.d. or current. This may be in the form of a step-function voltage or the act of connecting a sinusoidal source to a load or the act of altering, removing or adding a component to a network. All these variations will produce transient changes in the voltages and currents and after the transient effects a new steady state differing from that before the change will be maintained.

The most satisfactory way of determining the transient condition of a network is by the formulation of a differential equation which will describe its behaviour. The solution of this equation

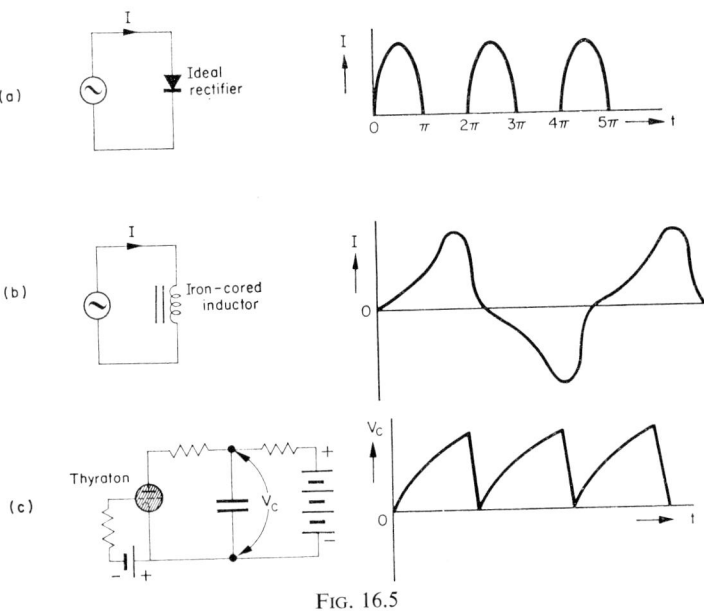

(a)
I
Ideal rectifier
I
0
π
2π
3π
4π
5π
t
(b)
I
Iron-cored inductor
I
0
(c)
Thyraton
V_C
+
−
+
V_C
0
t

FIG. 16.5

will often consist of two parts, one of which contains an exponential decay factor, so that that part eventually becomes zero, leaving the other part to represent the steady state of the network. Thus the differential equation gives us all the information we want about the operation of the network both during the period when transient conditions are important and afterwards.

Figure 16.6 shows a series L–R circuit where E is a function of e and t. The differential equation for this circuit is:

$$L\frac{di}{dt} + Ri = e$$

L
R
i
E

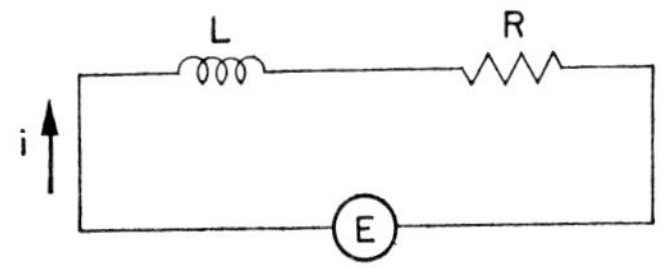

FIG. 16.6

The solution consists of a particular integral being any solution of the equation and a complementary function being a solution when $e = 0$. The latter is obviously independent of the driving voltage and must therefore be a characteristic of the network. For this equation:

$$A e^{-Rt/L} \text{ (where } A \text{ is a constant)}$$

is the complementary function, and it is clear that it decays with time, t, becoming zero after an infinite time (although from a practical point of view its value may become negligibly small quite rapidly).

If we wish to consider the effect of a step function applied to this circuit we must make $e = 0$ for $t < 0$ (i.e. for values of t before an arbitrary chosen time $t = 0$) $e = e_0$ for $t > 0$.

The equation thus becomes:

$$L \frac{di}{dt} + Ri = e_0$$

and

$$i = e_0/R$$

is a particular integral.

Thus the complete solution is:

$$i = \frac{e_0}{R} + A e^{-Rt/L}$$

The constant A is determined by conditions existing when $t = 0$. Just before the application of e_0 the current must be zero and just after the current must also be zero or else di/dt would be infinite requiring an infinite voltage across the inductor. Thus for $t = 0$, $i = 0$,

$$A = -e_0/R$$

$$\therefore i = \frac{e_0}{R} (1 - e^{Rt/L})$$

In a similar manner we have for the circuit in Fig. 16.7:

P.E.:
$$\frac{q}{C} + R \frac{dq}{dt} = e$$

where q is charge on C and $i = \dfrac{dq}{dt}$.

C.F. $A\, e^{-t/RC}$

Initially: $\begin{cases} t = 0 \\ q = 0 \end{cases}$

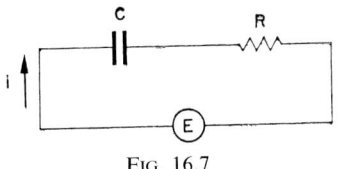

FIG. 16.7

Solution: $i = \dfrac{e_0}{R}\, e^{-t/RC}$

For circuit in Fig. 16.8:

P.E.: $L\dfrac{d^2q}{dt^2} + R\dfrac{dq}{dt} + \dfrac{q}{C} = 0$

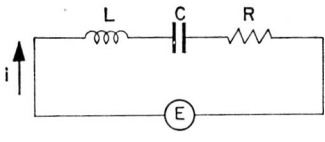

FIG. 16.8

C.F.: $A\, e^{-\alpha t} + B\, e^{-\beta t}$ for $R^2 > 4L/C$

where α, β are roots of $Lm^2 + Rm + (1/C) = 0$

and $e^{-\mu t}(A\cos \omega t + B\sin \omega t)$ for $R^2 < 4L/C$

where $\mu = -\dfrac{R}{2L}$

$$\omega = \sqrt{\left(\dfrac{1}{LC} - \dfrac{R^2}{4L^2}\right)}$$

and $A\, e^{-\mu t} + Bt\, e^{-\mu t}$ for $R^2 = 4L/C$

Initially:
$$\begin{cases} t = 0 \\ q = 0 \\ \dfrac{dq}{dt} = 0 \end{cases}$$

Solution for $R^2 < 4L/C$ is:

$$\frac{q}{C} = e_0[1 - e^{-\mu t}(\cos \omega t + (\mu/\omega)\sin \omega t)]$$

The extension of the above methods of solution for other waveforms of driving voltage and current is quite straightforward, E being made equal to the function describing the voltage. The resulting differential equation may, however, be more difficult to solve. With L–R and C–R networks an exponential decay factor is always found and for the L–C–R network we shall have either a damped sinusoidal oscillation or the sum of two decaying exponentials except when $R^2 = 4L/C$.

Example: A $1 \cdot 5\ \mathrm{k}\Omega$ resistor is connected in series with a 2 H choke and a voltage pulse is applied to the circuit by means of a generator having an output resistance of $500\ \Omega$ as shown in Fig. 16.9. The open-circuit voltage E provided by the generator is given by:

$$t < 0 \qquad E = 0$$
$$0 < t < 1\ \text{msec} \qquad E = 10\ \text{V}$$
$$1\ \text{msec} < t \qquad E = 0$$

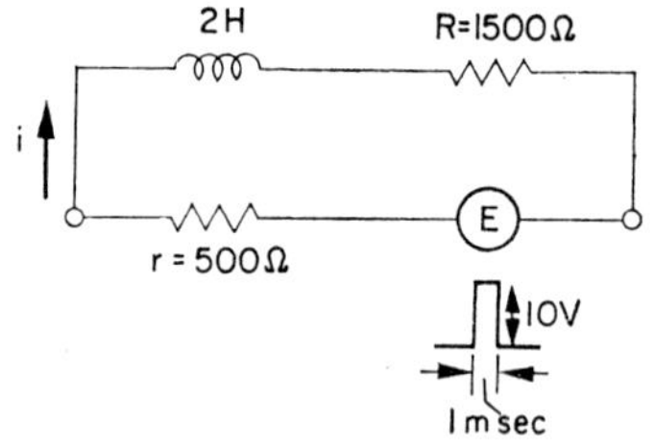

Fig. 16.9

Determine from first principles the voltage waveform V which appears across the resistor.

Solution:
$$E = L\frac{di}{dt} + (R+r)i$$

$$V = Ri$$

$E = E_0$ volts during pulse, zero before and after pulse

$$i = \frac{E}{R+r} + A \exp\left(-\frac{R+r}{L}t\right)$$

where A is determined from the fact that i through an inductor cannot change instantaneously.

Before pulse: $\quad\quad\quad i = 0$

During pulse: $\quad\quad\quad i = \frac{E_0}{R+r}\left[1 - \exp\left(-\frac{R+r}{L}t\right)\right]$

At end of pulse at $t = t_1$

$$i = \frac{E_0}{R+r}\left[1 - \exp\left(-\frac{R+r}{L}t_1\right)\right]\exp\left(-\frac{R+r}{L}(t-t_1)\right)$$

$$= \frac{E_0}{R+r}\left[\exp\left(\frac{R+r}{L}t_1\right) - 1\right]\exp\left(-\frac{R+r}{L}t\right)$$

$$\frac{L}{R+r} = 1 \text{ msec} \quad\quad t_1 = 1 \text{ msec}$$

$$\frac{R}{R+r} = 3/4$$

$$E_0 = 10 \text{ V}$$

$\therefore$ for $t < 0$ $\quad\quad\quad\quad V = 0$

$0 < t < 1 \text{ msec}$ $\quad\quad\quad V = 7 \cdot 5[1 - e^{-1000t}]$

$1 \text{ msec} < t$ $\quad\quad\quad\quad V = 7 \cdot 5[e-1]e^{-1000t}$

We can apply the above results to periodic waveforms, e.g. to a square wave as shown in Fig. 16.1(a), if we think of the wave as consisting of a positive step function lasting for time t followed by a negative one of the same amplitude and lasting for the same time. This will give us solutions for i which are shown graphically in Fig. 16.10(a), (b) and (c) for various values of L/R. In general the current wave form consequent on a periodic voltage change

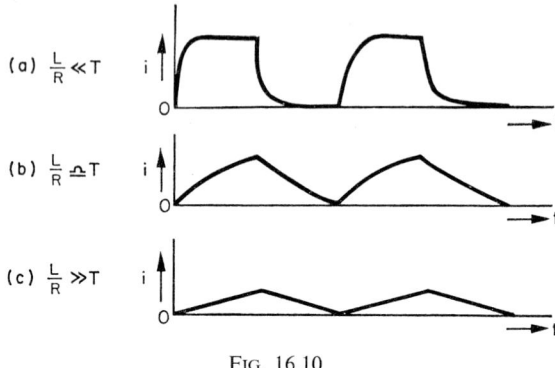

$$\text{(a)} \quad \tfrac{L}{R} \ll T \qquad i$$

$$\text{(b)} \quad \tfrac{L}{R} \cong T \qquad i$$

$$\text{(c)} \quad \tfrac{L}{R} \gg T \qquad i$$

Fig. 16.10

is not the same as the voltage waveform except in the case of sinusoidals. When the driving voltage is given by:

$$e = E_0 \cos \omega t$$

it can be shown that $i = I_0 \cos(\omega t - \alpha)$ is a particular integral of the differential equation if:

$$E_0 = I_0 \sqrt{(\omega^2 L^2 + R^2)}$$

and

$$\tan \alpha = \frac{\omega L}{R}$$

This is because

$$d/dt(\sin \omega t) = \omega \cos \omega t$$

and

$$d/dt(\cos \omega t) = -\omega \sin \omega t$$

thus giving a very convenient result for the relationship between voltages and currents in networks energized from sinusoidal supplies. It is because of this that the methods of solution described in earlier chapters are valid.

If we make $e = E_0 \cos \omega t$ in the differential equations for the circuits given in Fig. 16.6, 16.7 and 16.8, we shall get the following solutions for the current which may be compared with those obtained by vector methods given in an earlier chapter.

(a) $i = \dfrac{E}{\sqrt{(\omega^2 L^2 + R^2)}} \cos(\omega t - \theta)$ where $\tan \theta_1 = \dfrac{\omega L}{R}$

(b) $i = \dfrac{E_0}{\sqrt{\left(R^2 + \dfrac{1}{\omega^2 C^2}\right)}} \cos(\omega t + \theta_2)$ where $\tan \theta_2 = \dfrac{1}{\omega C R}$

(c) $i = \dfrac{E_0}{\sqrt{\left[\left(\omega L - \dfrac{1}{\omega C}\right) + R^2\right]}} \cos(\omega t - \theta_3)$

where $\tan \theta_3 = \dfrac{1}{R}\left(\omega L - \dfrac{1}{\omega C}\right)$

Example: Derive the differential equation for the current in the circuit of Fig. 16.11(a). If the circuit is initially quiescent and the generator produces a voltage step of height V_0 at $t = 0$, derive an expression for the current for $t > 0$ if $v^2 \ll L/C$.

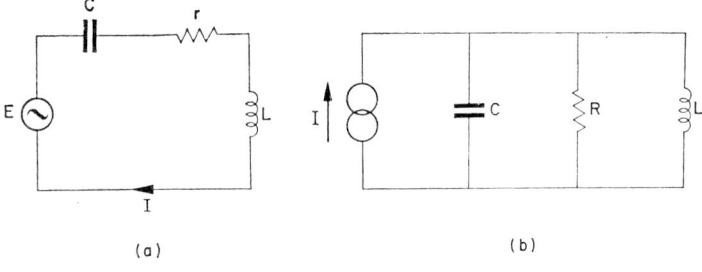

(a) (b)

FIG. 16.11

If a current step is applied to the circuit of Fig. 16.11(b) at $t = 0$ show how the solution to the first part of this example may be used to give the voltage across this circuit.

Solution:

For (a)
$$E = L\frac{di}{dt} + ir + \frac{1}{C}\int i\, dt$$

$$\therefore LC\frac{d^2 i}{dt^2} + rC\frac{di}{dt} + i = C\frac{dE}{dt}$$

C.F.:
$$i = e^{-\alpha t}(A\cos\beta t + B\sin\beta t)$$

where
$$\alpha = -\frac{r}{2L} \text{ and } \beta^2 = \frac{1}{LC} - \frac{r^2}{4L^2} \simeq \frac{1}{LC}$$

Particular integral $= 0$.

Initially:
$$\begin{cases} t = 0 \\ i = 0 \\ \dfrac{di}{dt} = \dfrac{V_0}{L} \end{cases}$$

$$\therefore A = 0 \text{ and } B = \frac{V_0}{\beta L} = V_0\sqrt{\frac{C}{L}}$$

$$\therefore i = V_0\sqrt{\left(\frac{C}{L}\right)}\, e^{-rt/2L}\sin\left(\frac{t}{\sqrt{(LC)}}\right)$$

For (b) it will be noticed that a current generator replaces a voltage generator and a parallel circuit replaces a series circuit. Now in a series circuit the current times the resistance or reactance gives voltage and in a parallel circuit a voltage times conductance or susceptance gives current. In a series circuit the impedance is the vector sum of the resistances and reactances and in a parallel circuit the admittance is the net sum of the conductances and susceptances. Thus the equations derived for the series circuit will be the same for the parallel circuit if we write:

$$\frac{1}{R}\text{ for } r \qquad C\text{ for } L \qquad v\text{ for } i$$

where v is the instantaneous voltage across the circuit.

Pairs of networks of the type considered in this problem are said to display the property of duality and are called duals.

Example: For Fig. 16.11(b)

$$I = C\frac{dv}{dt} + \frac{v}{R} + \frac{1}{L}\int v \, dt$$

Example: In Fig. 16.12 is shown a circuit containing a rapid acting switch S. After the circuit has reached a steady state the switch is closed and automatically re-opens permanently when the current through it becomes zero. Find for how long the

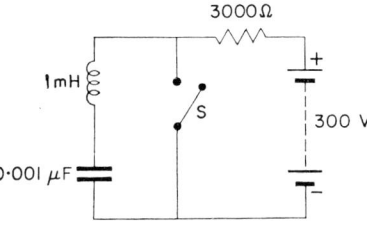

3000 Ω

1 mH

S

300 V

0·001 μF

FIG. 16.12

switch remains closed. Derive the differential equation satisfied by the p.d. across the capacitor after the switch has re-opened, together with the initial condition which the solution of this equation must satisfy.

Solution: Before closing, the charge q on C is CV and $dq/dt = 0$. After closing:

$$v = L\frac{d}{dt}\left(-V\frac{dv}{dt}\right)$$

$$\therefore \quad \frac{d^2v}{dt^2} + \frac{v}{LC} = 0$$

$$\therefore \quad v = V\cos\omega t \qquad \text{where } \omega = \frac{1}{\sqrt{(LC)}}$$

The current through $\qquad S = \dfrac{V}{R} - C\dfrac{dv}{dt}$

$$= \frac{V}{R} + \omega CV\sin\omega t$$

∴ switch opens when

$$\omega C V \sin \omega t + \frac{V}{R} = 0$$

$$\therefore \ \sin \omega t = -\frac{1}{\omega C T}$$

but

$$\omega = \frac{1}{\sqrt{(LC)}}$$

$$\therefore \ \sin \omega t = -\frac{1}{R} \sqrt{\frac{L}{C}}$$

$L = 10^{-3}$ H, $C = 10^{-9}$ F, $R = 3 \times 10^{3}\,\Omega$, $\omega = 10^{6}$ rad/sec

$$\therefore \ \sin \omega t = -1/3$$

$$\therefore \qquad t = \left(\pi + \sin^{-1}\frac{1}{3}\right) \mu\mathrm{sec}$$

$$= 3{\cdot}48 \ \mu\mathrm{sec}$$

Typical Examination Questions

1. A non-linear device is such that its output p.d. is given by

$$V_0 = 10(\sqrt{2}\sin \omega t + 0 \cdot 2 \sqrt{2}\sin pt)^2$$

If $\omega = 2\pi \times 10^6$ and $p = 2\pi \times 10^4$ rad/sec calculate the amplitudes of the components of each frequency present in the output waveform and the r.m.s. value of the output p.d.

> *Ans:* d.c. 10·4 V; a.c. at 2 Mc/s 10 V, at 20 kc/s 0·4 V, at 1·01 Mc/s 4 V and at 0·99 Mc/s 4 V; r.m.s. output 13·2 V

2. The input voltage to an amplifier has a square waveform, with equal "on" and "off" periods and a fundamental frequency of 1 kc/s. In order to avoid undue distortion, it is desired to retain all harmonic components whose amplitudes exceed 10 per cent of the amplitude of the fundamental component. What range of frequencies must be allowed for the transmission through the amplifier of this input?

> *Ans:* 0–9 kc/s

3. (a) A generator of zero internal impedance produces an e.m.f. of peak value E and isosceles triangular waveform of fundamental frequency f. The generator is connected in series with a resonant circuit consisting of a capacitor in series with an inductor. If the resonant circuit is tuned to a frequency $5f$ and has a quality factor Q at this frequency, what will then be the approximate peak voltage across the capacitor? It may be assumed that Q exceeds 100.

> *Ans:* $\dfrac{8EQ}{25\pi^2}$

(b) What would be the voltage if Q were the quality factor of the inductor alone and the capacitor had a power factor δ, both at frequency $5f$?

> *Ans:* $\dfrac{8EQ}{25\pi^2(Q\delta + 1)}$

4. In the circuits shown in Fig. 1 the switches are as shown for $t < t_0$. At $t = t_0$, they are switched to the alternative position. All the circuit elements are ideal. Obtain the differential equation applying in each case, when $t > t_0$; and hence the time constant (T.C.) in terms of L or C, and R. Give also the initial condition to be satisfied by the solution of the differential equation when $t = t_0$.

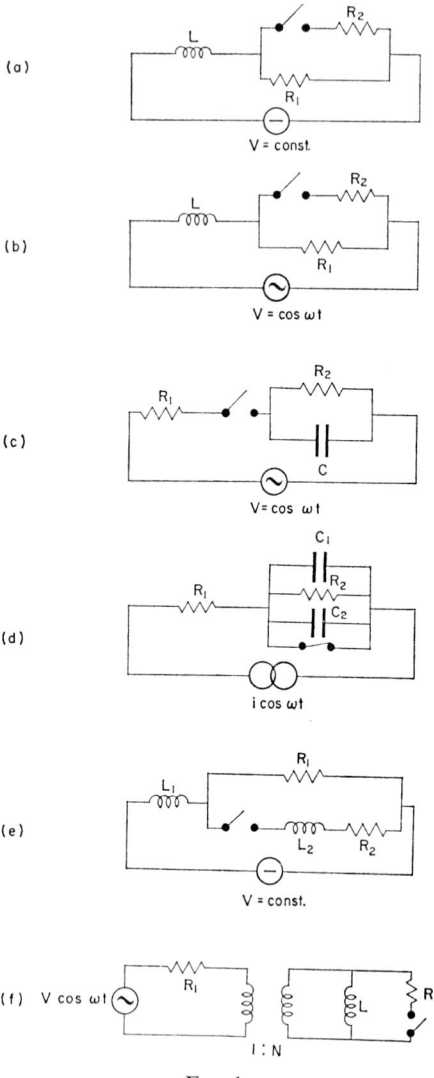

(a)

(b)

(c)

(d)

(e)

(f)

Fig. 1

Ans:

(a) T.C. $= L\left(\dfrac{R_1 + R_2}{R_1 R_2}\right)$ $i = \dfrac{v_0}{R_1}$ at $t = t_0$.

(b) T.C. $= L\left(\dfrac{R_1 + R_2}{R_1 R_2}\right)$ $i = v_0 \cos\left[\omega t_0 - \tan^{-1}\left(\omega L/R_1\right)\right]$
$$[R_1^2 + \omega^2 L^2] \text{ at } t = t_0.$$

(c) T.C. $= C\left(\dfrac{R_1 R_2}{R_1 + R_2}\right)$ $v = 0$ at $t = t_0$.

(d) T.C. $= (C_1 + C_2)R$ $v = 0$ at $t = t_0$.

(e) T.C. are $2L_1 L_2 / \{R_1(L_1 + L_2) + R_2 L_1 \pm \sqrt{[(R_1(L_1 + L_2) + R_2 L_1)^2}$
$$\overline{-4L_1 L_2 R_1 R_2]\}} \text{ at } t = t_0$$

i in $L_2 = 0$ $\dfrac{di}{dt}$ in $L_2 = v_0/L_2$.

(f) T.C. $= L\,\dfrac{N^2 R_1 + R_2}{N^2 R_1 R_2}$ $i = N v_0 \cos\left[\omega t_0 - \tan^{-1}\left(\omega L/N^2 R_1\right)\right]/$
$$\sqrt{[N^4 R_1 + \omega^2 L^2]} \text{ at } t = t_0$$

CHAPTER 17

Valves and Transistors as Circuit Elements

THERE are three distinct types of problem concerning the behaviour of triodes in circuits.

(a) Those in which the a.c. signals move the instantaneous operating point of the valve through such a wide range of values that it is not possible to consider the characteristics as linear. In such cases, superposition and Thévenin's theorem break down.

(b) Problems in which, although the a.c. signals are small, the d.c. operating point is located in a non-linear part of the characteristic such as in modulation and detection circuits.

(c) Operation in the linear part of the characteristic with signals small enough not to drive the valve beyond the limits of linearity.

We shall confine ourselves to the latter type of operation. A typical circuit for linear operation is given in Fig. 17.1(a), where

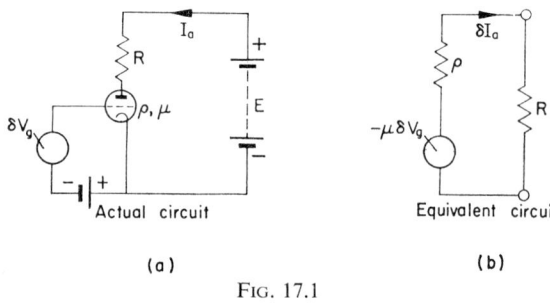

(a)   (b)

FIG. 17.1

the arrow shows the direction of conventional current flow, which is opposite to the direction in which the electrons actually move.

Then, if V_g increases by δV_g

$$\delta I_a = \frac{1}{\rho}[\delta V_a + \mu \delta V_g]$$

$$V_a = E - R I_a \qquad (E \text{ is const.})$$

$$\therefore \ \delta V_a = -R \delta I_a$$

and

$$\delta I_a = \frac{1}{\rho}[-R\delta I_a + \mu \delta V_g] = \frac{-R\delta I_a}{\rho} + \frac{\mu \delta V_g}{\rho}$$

$$\therefore \ \delta I_a\left(1 + \frac{R}{\rho}\right) = \frac{\mu \delta V_g}{\rho}$$

$$\delta I_a\left(\frac{\rho + R}{\rho}\right) = \frac{\mu \delta V_g}{\rho}$$

or

$$\delta I_a = \frac{\mu \delta V_g}{R + \rho} \quad \text{and} \quad \delta V_a = -\mu \delta V_g\left(\frac{R}{R + \rho}\right)$$

Now, we are usually only interested in changes of V_g, usually sinusoidal ones, and the corresponding changes in I_a. The changes are correctly represented by the *equivalent* circuit shown in Fig. 17.1(b) in which the valve is replaced by an equivalent generator giving an open-circuit e.m.f. $-\mu \delta V_g$ and having internal resistance ρ. The negative sign now associated with $\mu \delta V_g$ shows that the *change* in anode voltage is 180° out of phase with the *change* in grid voltage.

The above reasoning is valid not only for small changes in the level of V_g but also for small alternating voltages superposed on the mean d.c. value of V_g. Thus, if δV_g is of the form

$$V_g \sin \omega t$$

then

$$\delta I_a = \frac{\mu V_g \sin \omega t}{R + \rho}$$

If the anode load is more complex than a pure resistance, e.g. an impedance Z, the equivalent circuit will have Z in place of the R shown in Fig. 17.1(b).

In this case the driving voltage is still 180° out of phase with V_g, but the output voltage, which depends on Z and I, may have any phase whatsoever.

To summarize:

(a) Replace triode by equivalent voltage generator and all normal linear circuit theory applies.

(b) But this is only valid for small a.c. signals for which μ and ρ remain constant.

(c) Also, for small a.c. signals it is valid to consider a.c. and d.c. voltages and currents separately. The d.c. voltages V_a, V_g with R set the operations point and thus specify μ and ρ.

The a.c. quantities then follow from the equivalent circuit without reference to the d.c. currents or voltages, provided that the internal resistance of battery E is so low that no a.c. voltage is developed across it. This is ensured by placing a large capacitor across the battery, so that the + battery terminal is at earth potential for a.c.

Note that by small signal we merely mean one which does not drive the valve out of the linear range. A small signal for a transmitting valve might be several hundred volts.

Amplification

For resistive load

$$i_a = \frac{-\mu V_g}{R + \rho}$$

$$\therefore \; V_a = i_a R = \frac{-\mu V_g R}{R + \rho}$$

$$\text{Voltage amplification} = \frac{V_a}{V_g} = \frac{-\mu R}{R + \rho}$$

Thus, amplification $\to \mu$ as $\rho/R \to 0$.

In actual fact, most modern small signal amplifiers use pentode valves rather than triodes. As the name implies, the pentode has five electrodes, three of which are grids between anode and cathode. G_1 is nearest to the cathode and acts similarly to the triode grid. It is called the control grid. G_2 is held at a positive potential in the vicinity of the anode potential and is called the screen grid, whilst G_3, nearest the anode, is held at cathode potential or at a slightly negative potential and is called the suppressor grid.

In the pentode, the current through the valve is determined almost entirely by the characteristics of, and the potentials applied to, the triode formed by C, G_1 and G_2. In particular the value of V_a has very little influence on the current in the valve so that

$$\frac{\partial I_a}{\partial V_a} \to 0 \quad \text{and} \quad \rho \to \infty$$

The pentode g_m is that of the above-mentioned triode and we therefore deduce that

$$\mu = g_m \, \rho$$

where g_m is the mutual conductance, is also very large for the pentode.

Characteristics

The I_a, V_g characteristic is similar to that of a triode but the I_a, V_a characteristic is quite different. It is of the form shown in Fig. 17.2.

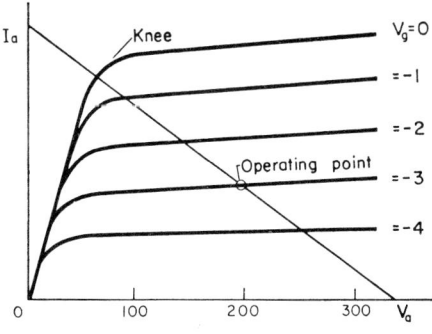

Pentode characteristic curves

FIG. 17.2

The knee voltage is quite low, between 50 and 100 V. The reason for the appearance of the knee is that since the suppressor potential is zero or negative, when the anode potential is low the electron velocities in the anode region are low and the space-charge density is high. Thus, the space-charge can depress the potential between G_s and the anode to below zero and turn some of the electrons back. When V_a is low, some electrons go back but when it is high all reach the anode.

The function of the suppressor is to ensure that secondary electrons from the anode cannot return to the screen, even when $V_a < V_s$. This is the defect of the screen grid valve, now obsolescent.

The load line is drawn and the operating point fixed just as with the triode. But a more convenient equivalent circuit is possible, owing to the high value of ρ. Since ρ is so high, the pentode is practically speaking a constant current generator and it is best to represent it in current generator form as in Fig. 17.3.

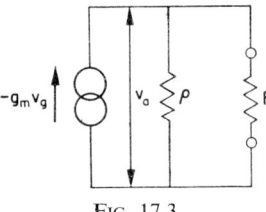

FIG. 17.3

Then

$$V_a = -g_m V_g \frac{\rho R}{\rho + R}$$

Very often $\rho \gg R$, so that

$$V_a \simeq -g_m V_g R$$

or

$$A = \frac{V_a}{V_g} = -g_m R.$$

This is convenient as the precise value of ρ is now unimportant, as long as it is $\gg R$.

In a manner similar to that given above for valves the analysis of circuits containing transistors can be made with sufficient accuracy for most purposes by the application of the principles already developed provided that a suitable equivalent circuit, or model, can be devised from consideration of the electrical characteristics of the device. There are two main types of transistor, the bipolar transistor and the field effect transistor (or f.e.t. for short); we shall deal with the bipolar first.

Bipolar Transistor

Consider the n–p–n sandwich of semiconductor shown in Fig. 17.4. This contains two p–n junctions facing outwards. If the *collector* region is more positive than either of the other regions, then the collector to base, n–p junction is reverse biased and so only a small leakage current flows (which depends on temperature).

If the centre region, *the base*, is made more positive than the *emitter*, then the base-emitter, p–n junction will conduct as in an ordinary diode.

The emitter region is heavily doped. Electrons are the majority carriers in n-material and these flow to the base emitter junction and into the base. The base is lightly doped and so only a few holes, the majority carriers in p-material, flow from base to emitter. If the base region is very thin, only a few electrons from the emitter flow to the base connection or are neutralized by holes and the majority flow through the base to the more positive collector.

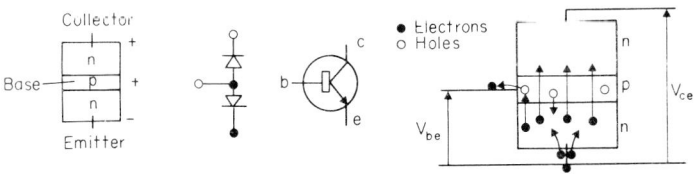

FIG. 17.4

The collector current is thus many times larger than the base current. The factor, I_C/I_B will be large when

(a) the base doping is much less than the emitter,

(b) the base region is very thin.

Normally it is of the order of 100 and it may be nearly constant over a wide range of current. It is called the Beta, h_{FE}, or Static Common-Emitter Current Gain of the transistor.

Figure 17.5 shows a transistor supplied by current sources I_1 and I_2, of adjustable strength. Source I_1 drives current into the base terminal b, I_2 supplies the collector terminal c, and the third terminal of the transistor, the emitter e, is common to both the input and the output circuits and carries $I_1 + I_2$. If the voltages across the terminals V_{be} and V_{ce} are then measured for a range of values of I_1 and I_2, the volt-amp characteristics of Fig. 17.6(a) and (b) result, which are typical of a small transistor.

It is most convenient to think of a transistor as a current-operated device, so that the measurements are plotted in the form

$$V_{be} = f_1(I_b, I_c) \qquad (I_c \text{ as parameter}) \qquad (1)$$

$$V_{ce} = f_2(I_c, I_b) \qquad (I_b \text{ as parameter}) \qquad (2)$$

This way of connecting a transistor is the most usual, since it gives the highest power gain. It is called the *common-emitter* circuit since the emitter is common to both the input and output circuits. Other configurations are also used; the collector may be the common terminal or the base may be, giving the emitter-follower and the common-base arrangements respectively. A parallel is the thermionic triode where the common-cathode is the usual connection and the cathode-follower and the grounded-grid circuits are used for special purposes.

If we are primarily interested in the response of a transistor to small changes in the current variables, that is to small signal variations, then (1) and (2) can be differentiated in the usual way

$$\delta V_{be} = \frac{\partial V_{be}}{\partial I_b} \, \delta I_b + \frac{\partial V_{be}}{\partial I_c} \, \delta I_c$$

$$\delta V_{ce} = \frac{\partial V_{ce}}{\partial I_c} \, \delta I_c + \frac{\partial V_{ce}}{\partial I_b} \, \delta I_b$$

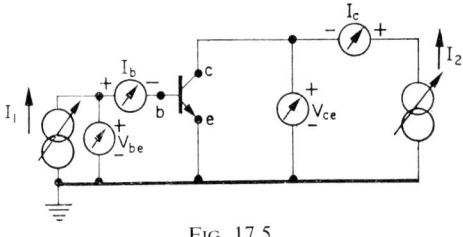

FIG. 17.5

If the signals are small enough, the partial differential co-efficients become constants, and

$$\delta V_{be} = r_{11}\,\delta I_b + r_{12}\,\delta I_c \qquad (3)$$

$$\delta V_{ce} = r_{21}\,\delta I_b + r_{22}\,\delta I_c \qquad (4)$$

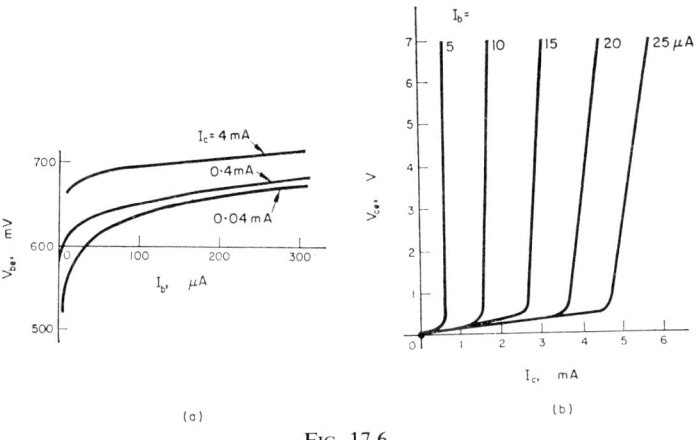

(a)

(b)

FIG. 17.6

where the coefficients are now the resistive form of the so-called z parameters (never given in manufacturers' data). They are also a set of small signal equations governing any linear four terminal network.

We could, in principle, get the coefficients r_{11}, etc., by measurements say of $\delta V_{be}/\delta I_b$ with I_c constant, so that $\delta I_c = 0$. If the resistance parameters were known, then the small signal performance of the transistor circuit could be calculated for any given source and load impedance.

One can see from a derivation of this sort how the small signal behaviour might be deduced from the device characteristics. However many people find it more helpful to use a circuit model, rather than a four-terminal box, even if the box has just the same small signal characteristics. Such a circuit model is in fact just as artificial and the values assigned to its parts have to be found by circuit measurements.

Ideal Small-Signal Current Amplifiers

Let us approach a circuit model from a different point of view. We want a device which will amplify small current changes: if δI_b = a change of input current to the device then we would like $\beta \delta I_b$ to be the ensuing change in its output current, with β large and constant.

A perfect small signal current amplifier would have zero input resistance (or some of the incoming current would be diverted) and infinite output impedance, or some of the output current would be by-passed in the output resistance. We can expect real devices to fall short of this ideal, and could represent them as in Fig. 17.7.

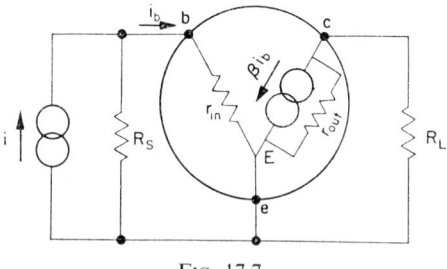

Fig. 17.7

This shows a real small signal current source in parallel with its internal resistance R_s, the amplifier (inside the box), and an external load R_L. Since we are only concerned with small changes of current any steady currents, by superposition, are irrelevant and are not shown. For simplicity i, i_b, etc., have been written for δI, δI_b, etc.: it is understood that we are concerned with small variations of the standing currents, i.e. with small signals, and of course one particular case of this is $\delta I = i \sin \omega t$, small sinusoidal signals, or δI may have the representation $i e^{j\omega t}$.

The input resistance of the modified amplifier is not zero but r_{in}, its output resistance is r_{out}, and not ∞.

So

$$i_b = \frac{R_s}{R_s + r_{\text{in}}} \times i$$

and is less than i; and

$$i_L = \frac{r_{\text{out}}}{r_{\text{out}} + R_L} \times \beta i_b$$

and so is less than βi_b. Thus the existence of a finite input and output resistance has reduced the amplification compared with that of an ideal current amplifier.

But the perfect amplifier has another virtue: the output circuit does not react on the input circuit. The model of Fig. 17.7 has this ideal property too, since load current flowing in eE causes no IR drop because the branch eE, in this model, has no resistance. But real amplifiers do have feedback from the output circuit to the input, and the simplest way to make our equivalent circuit show this effect is to insert a resistor in eE, which is common to both input and output circuits. Thus the model of Fig. 17.8 has finite r_{in}, r_{out}, and feedback from output to input

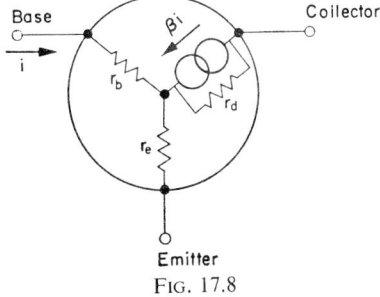

Fig. 17.8

circuit: it suffers from all the defects of a real current amplifier; and can give the same variational circuit behaviour as a transistor if r_b, r_e, r_d, β are given the correct values. Thus we could obtain formulae for r_b, r_e, r_d and β in terms of the resistance parameters but the formulae are not simple. The circuit model of Fig. 17.8 is called the *equivalent tee* small signal circuit. It is worth emphasizing that all the quantities r_b, r_e, r_d, β represent circuit quantities and follow from but are not *directly* related to transistor physics. Their values are always deduced from circuit measurements.

Small Signal Performance of the Common Emitter Amplifier

Figure 17.9 shows a signal source driving a transistor, R_L is the load. Any real source can be represented either by i in parallel with R_s, or e in series with R_s, with $e = iR_s$. The current-operated source is used in Fig. 17.9. We will calculate (1) the input impedance which the transistor presents to the source, (2) the output resistance seen by the load, (3) the current gain. From these the

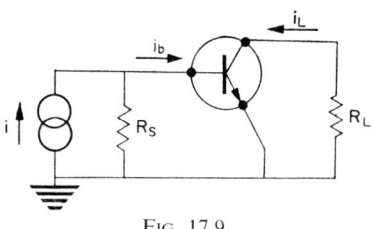

FIG. 17.9

voltage and power gains can easily be obtained. By superposition, the d.c. currents and voltages and the d.c. energy sources are irrelevant to the variational calculation, and are not shown.

(a) *Input resistance.* Drive the transistor by an ideal current source i_b, as shown in Fig. 17.10. Use superposition and calculate the effects of i_b and βi_b separately.

$r_{\text{in}} = \dfrac{v}{i_b}$, and $v =$ p.d. across $r_b +$ p.d. across r_e

$$= i_b r_b + r_e \left(i_b \underbrace{\frac{r_d + R_L}{(r_d + R_L) + r_e}}_{(A)} + \beta i_b \underbrace{\frac{r_d}{r_d + (R_L + r_e)}}_{(B)} \right)$$

(A)   (B)

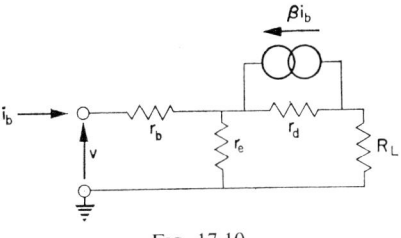

FIG. 17.10

The term (A) is the fraction of i_b that flows in r_e, and the term (B) is the fraction of βi_b that flows in r_e. Since i_b is an ideal current source, βi_b cannot drive current through r_b, for r_b is in series with an infinite resistance.

$$\therefore \ r_{\text{in}} = \frac{v}{i_b} = r_b + r_e \cdot \frac{\beta r_d + r_d + R_L}{r_d + R_L + r_e}$$

$$= r_b + r_e \frac{\beta + 1 + (R_L/r_d)}{1 + (R_L/r_d) + (r_e/r_d)}$$

Now R_L/r_d is seldom > 0.5, β is between 30 and 300 and is not known within a factor of 3. Hence a sensible approximation for the term

$$[\beta + 1 + (R_L/r_d)] \text{ is } \beta$$

r_e/r_d is usually $1/1000$, $\ll 1$.

$$r_{\text{in}} \simeq r_b + \frac{\beta r_e}{1 + (R_L/r_d)} \tag{5}$$

Notice the effect of internal feedback, r_{in} depends on the output circuit because of the term (R_L/r_d).

(b) *Current amplification i_L/i_b.* Again evaluate the effects of i_b and βi_b separately. Due to βi_b, a current

$$\beta i_b \cdot \frac{r_d}{r_d + (r_e + R_L)}$$

flows in R_L in the direction shown, and due to i_b, current

$$-i_b \frac{r_d}{r_d + (r_e + R_L)}$$

flows in the same direction.

$$\therefore \ i_L/i_b = \frac{\beta r_d - r_e}{r_d + R_L + r_e} = \frac{\beta - (r_e/r_d)}{1 + (R_L/r_d) + (r_e/r_d)} \simeq \frac{\beta}{1 + (R_L/r_d)} \qquad (6)$$

(c) *The output circuit and output resistance.* We can represent the output side of the transistor by a single current source i_t in parallel with a resistance r_t, since for the linear small-signal behaviour Thévenin's theorem is applicable. This is especially useful if the transistor is part of a cascade. By "output resistance" we mean just the Thévenin resistance r_t in Fig. 17.11.

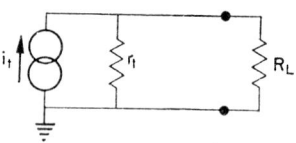

Fig. 17.11

Notice that if $R_L \rightarrow 0$, all of i_t flows in the short circuit, none flows in r_t. In equation (6) the current flowing into the transistor collector when R_L is zero is βi_b, so that the magnitude of i_t is $|\beta i_b|$.

But now we consider the Thévenin source current to flow outwards, away from the collector, so with opposite sign:

$$i_t = -\beta i_b \qquad (7)$$

We can calculate the output resistance r_t from the ratio

$$\frac{e_t}{i_t} = \frac{\text{output voltage when } R_L = \infty}{\text{output current when } R_L = 0}$$

When $R_L = \infty$, then $r_{\text{in}} = r_b + r_e$, from the exact form of equation (5) when $R_L = \infty$. In which case

$$i_b(R_L = \infty) = \frac{iR_s}{R_s + r_b + r_e}$$

From Fig. 17.12.

$$e_t(R_L = \infty) = -\beta i_b r_d + i_b r_e \qquad \text{(note the signs)}$$

$$= -i(\beta r_d - r_e \frac{R_s}{R_s + r_b + r_e}$$

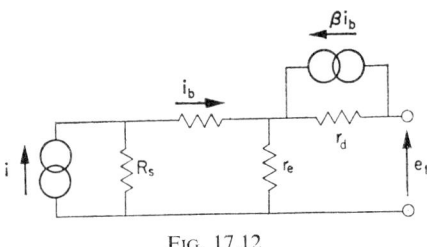

FIG. 17.12

When $R_L = 0$, then, from the exact form of (5),

$$r_{\text{in}} = r_b + \frac{(\beta + 1)r_e}{1 + (r_e/r_d)} = r_b + (\beta + 1)r_e$$

very closely.

$$\therefore i_b(R_L = 0) = \frac{iR_s}{R_s + r_b + (\beta + 1)r_e}$$

$\therefore i_L$, into the collector, $= \beta i_b$ (by (6) when $R_L = 0$), and i_t, which is out of the collector, is

$$i_t = -\beta i_b$$

$$= -\frac{i\beta R_s}{R_s + r_b + (\beta + 1)r_e}$$

$$\therefore r_t = \frac{e_t}{i_t} = \frac{(\beta r_d - r_e)R_s}{\beta R_s} \cdot \frac{R_s + r_b(\beta + 1)r_e}{R_s + r_b + r_e}$$

$$= \left(r_d - \frac{r_e}{\beta}\right)\left(1 + \frac{\beta r_e}{R_s + r_b + r_e}\right)$$

$$\simeq r_d[1 + \beta r_e/(R_s + r_b + r_e)] \qquad (8)$$

Return to the overall circuit now, Fig. 17.13.

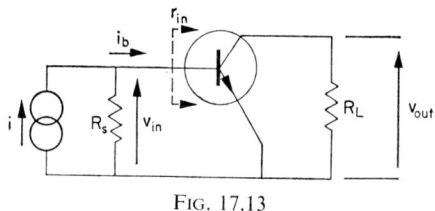

FIG. 17.13

$$i_b = i\,\frac{R_s}{R_s + r_{\text{in}}} \qquad (r_{\text{in}} \text{ is given by (5)})$$

$$v_{\text{in}} = i_b r_{\text{in}}$$

$$v_{\text{out}} = -R_L i_L \qquad (i_L \text{ is given by (6)})$$

$$\frac{\text{a.c. power in } R_L}{\text{a.c. power supplied to the transistor}} = \frac{v_{\text{out}}\,i_L}{v_{\text{in}}\,i_b}$$

Example: Let $\beta = 100$, $r_e = 10\,\Omega$, $r_b = 1\,\text{k}\Omega$, $r_d = 20\,\text{k}\Omega$, $R_L = 10\,\text{k}\Omega$, and the signal source be $10\,\mu\text{A}$ shunted by $10\,\text{k}\Omega$ (a preceding transistor stage perhaps):

$$r_{\text{in}} = 1000 + \frac{100 \times 10}{1 + 0 \cdot 5} = 1 \cdot 67\,\text{k}\Omega$$

$$i_b = 10 \times \frac{10}{10 + 1 \cdot 67} = 8 \cdot 57\,\mu\text{A}$$

$$i_L = \frac{100}{1+0\cdot5}\, i_b = 66\cdot7 i_b = 0\cdot57 \text{ mA}$$

$$v_{\text{in}} = i_b r_{\text{in}} = 14\cdot3 \text{ mV} \qquad v_{\text{out}} = -i_L R_L = -5\cdot7 \text{ V}$$

$$v_{\text{out}}/v_{\text{in}} = \text{voltage gain} = -395$$

$$i_{\text{out}}/i_{\text{in}} = \text{current gain} = 66\cdot7$$

$$\frac{\text{a.c. power out}}{\text{a.c. power in}} = A_i\, A_v = 66\cdot7 \times 395 = 26{,}300$$

If this transistor were to drive another, then

$$r_o = \left(1\Big/\frac{100\times10}{10+1000+10{,}000}\right) 20{,}000 = 21\cdot8 \text{ k}\Omega$$

and the next stage would see R_L and r_o in parallel, or about 6·8 kΩ, as the impedance of its drive. And the next stage source current i_t would be $-\beta i_b$ or $-0\cdot857$ mA.

The h-parameters will now be considered. These are mentioned because so much data is given in terms of them. Recall that we arrived at the common-emitter circuit by finding a model in which the ideal current amplifier was degraded to have the imperfections of actual amplifiers. But the feedback could be

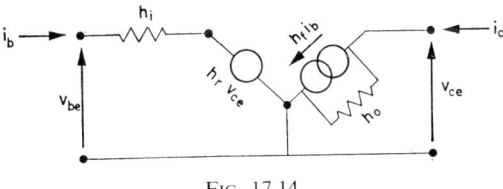

FIG. 17.14

modelled in another way, and in Fig. 17.14 a circuit which uses an explicit feedback generator is shown.

By inspection

$$V_{be} = h_i i_b + h_r v_{ce} \qquad (h_i = \text{resistance})$$

$$i_c = h_f i_b + h_o v_{ce} \qquad (h_o = \text{admittance})$$

These "h", or "hybrid" parameters lead to another set of four terminal network equations.

To get the tee parameters in terms of the h ones we shall compare circuits (a) and (b) in Fig. 17.15:

(1) Compare output currents when c is shorted to earth

$$i_c = h_f i_b = \beta i_b$$

(2) Compare input resistance when c is shorted to earth

$$r_{in} = h_i = r_b + \beta r_e$$

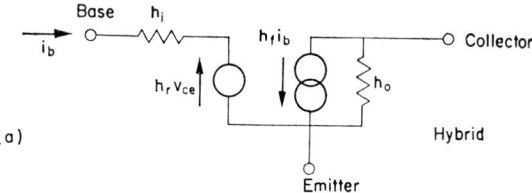

(a)   Hybrid

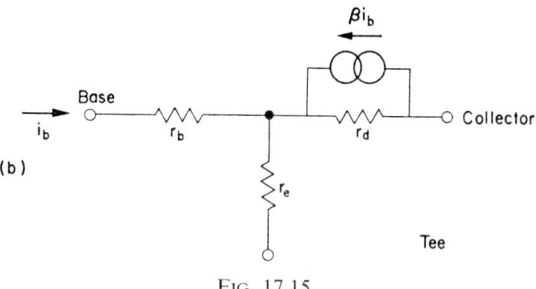

(b)   Tee

FIG. 17.15

(3) Compare output resistance when the input is open and $i_b = 0$.

$$\frac{1}{h_o} = r_d + r_e$$

Feedback voltage transmission with the input open,

$$h_r v_{ce} = \frac{r_e}{r_e + r_d} v_{ce}$$

$$\therefore\ h_i = r_b + r_e(1+\beta) \quad \text{and} \quad \beta = h_f$$

$$h_r = r_e/r_d \qquad\qquad r_d = 1/h_o$$

$$h_f = \beta \qquad\qquad r_e = h_r/h_o$$

$$h_o = 1/r_d \qquad\qquad r_b = h_i - \frac{h_f h_r}{h_o}$$

There are two kinds of transistors with opposite polarities. We have used "n–p–n" transistors throughout, in which the currents I_b and I_c flow inward for correct operation, and both base and collector are held positive with respect to the emitter. Thus if the standard sign convention, which applies equally to valves and to four terminal networks, is used (Fig. 17.16) measurements of I_b, I_c, V_{be}, V_{ce} give positive numbers with n–p–n transistors.

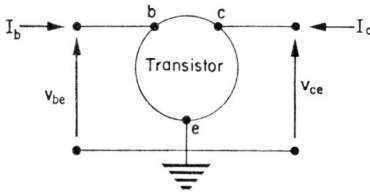

FIG. 17.16

The other kind of transistor, the "p–n–p" type, must have its base and collector held negative with respect to its emitter, and base and collector currents flow outwards. So that with the sign convention of Fig. 17.16, measurements of I_b, I_c, V_{be} and V_{ce} all give numbers which are negative.

The symbols for p–n–p and n–p–n transistors differ in that the emitter arrow points inward for p–n–p, outward for n–p–n; it points in the direction of current flow in the emitter. If the convention that positive voltage lines are drawn at the top is used (it often isn't) then we get the symbols of Fig. 17.17.

The electrical sizes of transistors vary from small ones which can operate with standing collector currents of $10\,\mu\text{A}$, to large power types with quiescent collector currents of amps and which

handle current swings of tens of amps. The latter can dissipate
50 W at their collectors. Maximum voltages are seldom greater
than 100 V. The circuits discussed here are not applicable at
higher frequencies, when more elaborate equivalent circuits must

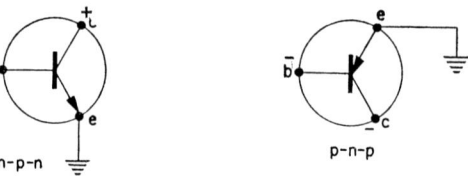

FIG. 17.17

be used. The gain of transistors falls off with frequency, but f_T,
the frequency at which the common-emitter power gain falls to
unity, may be 50 Mc/s in some small types, and useful power
gains at up to 500 Mc/s can be got in the common-base circuit.

Transistors are made either of germanium or of silicon. The
latter have much lower leakage currents, so that variation of
I_{co} with temperature may be unimportant; they are also electrically more robust.

Comparative Table of Symbols

These are far from standardized and the table below gives a
list of alternatives:

CE Current amplification factor	$\beta, b, h_f, h_{fe}, a', a_{cb}$
CB Current amplification factor	a, a, h_f, h_{fb}
Emitter-open leakage current	I_{CO}, I_{CBO}
Base-open leakage current	I'_{CO}, I_{CEO}
CE collector branch resistance	$r_d, R_d, r_c(1-a)$
CB collector branch resistance	$r_c, r_{cc}, r_d(1+\beta), R_c$
Equivalent tee emitter resistance	r_e, R_e

Emitter junction diode forward resistance r_e ($\neq r_e$)
Equivalent tee base resistance r_b, R_b
Extrinsic, or base spreading, resistance r_{bb} ($\neq r_b$)

There is a low-frequency "physical" model of the common base transistor in which each element has a physical interpretation, and the values of each element can be calculated (in principle) in terms of physical processes. The model has two active dependent generators and is too complicated for use in circuit analysis. It is transformed, by circuit transformations, into a one-generator CB equivalent tee. The elements of the circuit CB tee are all different, in principle, from those of the physical CB tee, but the differences in a, and in r_c are parts in a thousand. But r_e and r_{bb} in the physical tee are appreciably different from the r_e and the r_b used in the circuit tees. The one-generator circuit CB tee can itself be transformed, again by a circuit transformation, into the one-generator circuit CE tee.

Field Effect Transistor

The f.e.t. has the following features: A very high input impedance, conduction by majority carriers, low noise, low drift and good frequency response. These features make it possible to improve the performance of many circuits based on junction transistors and enable the design of circuits which were not feasible without f.e.t.s.

Ohmic contacts are made on two opposite faces of a Si bar which is called the *channel* (n-channel if made with n-type material and p-channel if made with p-type). The contacts are called the *source* and *drain* respectively. Electrons are injected into the channel at the source and collected at the drain, i.e. a current, I_D, flows from drain to source when a battery is connected as in Fig. 17.18(a). The magnitude of I_D, which is a majority carrier current, depends upon the conductivity of the channel and the voltage V_{DS}.

Two p–n junctions are grown on two opposite transverse faces of the channel and are known as *gates*. (Fig. 17.18(b).) They can be used to control the magnitude of the current I_D, because, if they are reverse-biased with respect to the source, a depletion layer will extend into the channel and decrease its effective width.

Recall that, as explained for the p–n junction diode, there are very few carriers in the depletion layer and its resistance is very high.

The channel resistance would, therefore, be increased and the drain current limited.

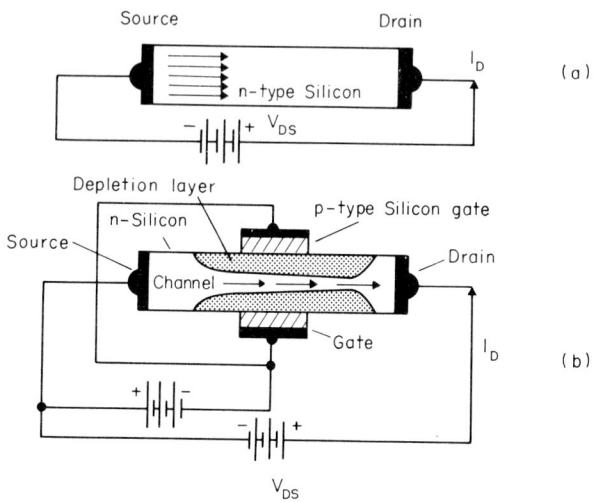

Fig. 17.18

Consider the gate voltage to be zero w.r.t. the source. Drain current flows and causes a voltage drop along the length of the channel such that there is a small reverse bias, between gate and source, at the source end, and a much greater reverse bias at the drain end. A wedge-shaped depletion region forms as shown in Fig. 17.18(b). If I_D is increased the width of this depletion layer increases, the channel is narrowed and its resistance increases. An equilibrium state is reached when no further increase of current is possible because the channel is almost closed at the drain end. At this stage saturated current flows and the channel resistance is very high. Any further increase of V_{DS} and extension of the depletion regions is counter-balanced by the high field in the very narrow channel allowing saturation current to flow. At still higher drain-source voltages there is a gate-drain breakdown due to avalanche effects and f.e.t. action ceases.

The device characteristic with $V_G = O$ is shown in Fig. 17.19. The point at which the current saturates is known as *pinch-off*. Beyond this an almost constant current is obtained until the device breaks down. If the gate is made negative w.r.t. the source the whole process described for $V_G = O$ is repeated except that pinch-off occurs at a lower value of drain current. Thus, with V_G as a parameter a family of curves is obtained as shown in Fig. 17.20. The locus of the pinch-off points is known as the pinch-off curve and the region of operation to its left is known as the triode region while the region to its right is called the pinch-off or constant current region.

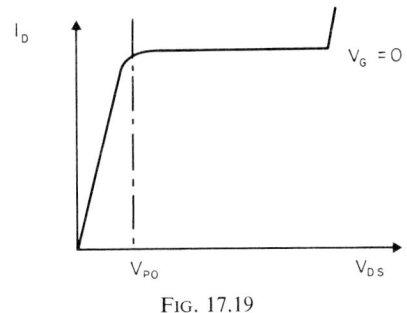

FIG. 17.19

The device described in the foregoing is said to work in the *depletion-mode*, i.e. a negative gate voltage is used to *decrease* channel width and produce pinch-off. It is possible to start with a device designed such that pinch-off is obtained for very low values of drain current with $V_G = O$. The gates of such a device are biased $+ve$ w.r.t. the source in order to open a wider channel so that pinch-off occurs at greater values of drain current. This device is said to work in the *enhancement mode*. It is also possible to construct devices capable of working in both modes so that V_G can be both $+ve$ and $-ve$ w.r.t. the source.

The description of the n-channel device applies equally to the p-channel type except that the conduction process is described in terms of hole movements.

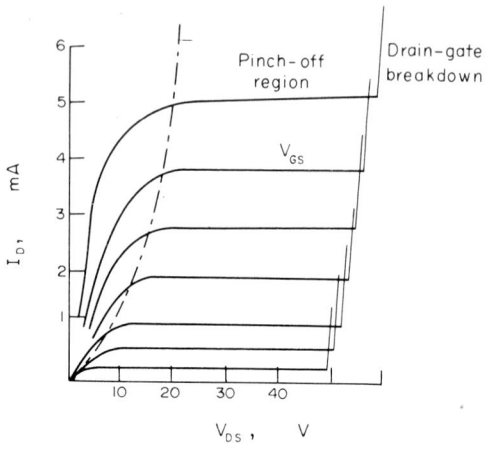

Fig. 17.20

The symbols for f.e.t.s. are shown in Fig. 17.21 and compared with those for transistors and valves.

Note that for an n-channel device the gate arrow points *inwards* and the drain is +*ve* while the gate is −*ve* w.r.t. the source.

For the p-channel device the arrow points *outwards*, the drain is −*ve* and the gate is +*ve*.

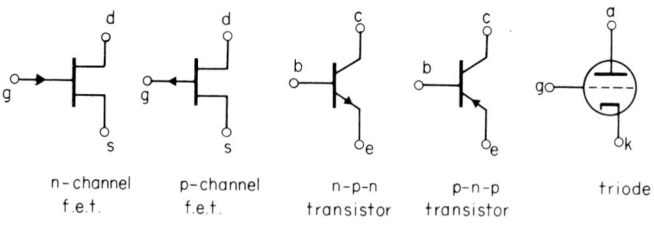

Fig. 17.21

The characteristics of the f.e.t. (Fig. 17.20) are pentode like beyond the pinch-off curve. The incremental model of the pentode, therefore, adequately describes the f.e.t. The input impedance of the f.e.t. is a reverse biased p–n junction which may be represented as a very large resistance together with a small capacitor in parallel. The enhancement-mode device has a low input impedance because the junction is forward biased and it will not be considered. (Variations of the f.e.t. known as MOSTS have very high input impedance in either mode).

Returning to the characteristics of Fig. 17.20 we define a mutual conductance g_m and a drain resistance r_d as;

$$g_m = \left(\frac{\Delta I_D}{\Delta V_G}\right) V_{DS}, \quad \gamma_d = \left(\frac{\Delta V_{DS}}{\Delta I_D}\right) V_G$$

The incremental model is shown in Fig. 17.22 for low and medium frequencies. For high frequencies it is modified as shown in Fig. 17.23 where gate-source and drain-source capacities are included.

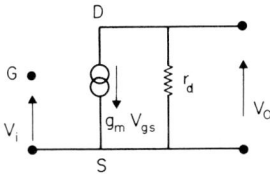

FIG. 17.22

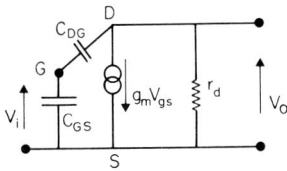

FIG. 17.23

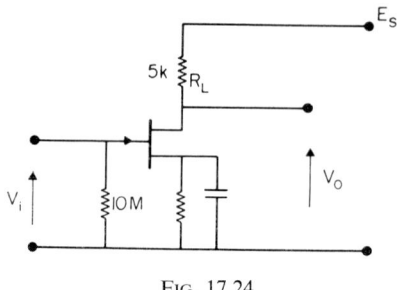

FIG. 17.24

Finally an amplifier circuit using a f.e.t. is shown in Fig. 17.24. The analysis for the gain, etc., follows the usual practice of drawing the incremental model and setting up the loop or nodal equations.

Typical Examination Questions

1. Explain the following terms:
(a) Unipolar devices; (b) Bipolar devices; (c) F.e.t.; (d) Channel, drain and source; (e) Emitter, base and collector; (f) MOST; (g) Pinch off.

2. The characteristics of a p-channel f.e.t. are shown in Fig. 1.
(a) Estimate the values of g_m and r_d for the operating point Q marked on the load line.
(b) Estimate the load resistance of a grounded source amplifier with the load line as in Fig. 1.
(c) Estimate the voltage gain for a 1 V pk-pk signal applied to the gate.
Ans: (a) 2 mA/V, 60 kΩ; (b) 2·5′kΩ; (c) 5.

3. The Mullard BC 107, Silicon npn transistor has the following typical working point:
$$V_{cw} = +10 \text{ V}, I_c = 25 \text{ mA}, I_b = 100 \,\mu\text{A}, V_{be} = +0\cdot7 \text{ V}.$$
At this working point, its small signal parameters are:
$$\beta - 250, r_d = 3\cdot3 \text{ k}\Omega, r_e = 1 \,\Omega, r_b = 750 \,\Omega.$$
(a) If the supply voltage is +20 V, what load resistor and base bias resistor will achieve the working point given in the simple common emitter amplifier of Fig. 2?
(b) Find the Current Gain, Input Resistance, Voltage Gain and Power Gain for the amplifier of Fig. 2. (Ignore the effect of C and of R_B in the circuit.)

(c) Graphical data of the BC 107 is given in Fig. 3. For a supply of $+20$ V, put a 400 Ω load line across the output characteristic and check that the working point at (a) is obtainable.

For an input swing of ± 50 μA about the standing base current of 100 μA, get graphically the voltage gain of the amplifier with a 400 Ω load.

Ans: (a) 400 Ω, 93 kΩ; (b) 223, 913 Ω, -92, 20500; (c) -80.

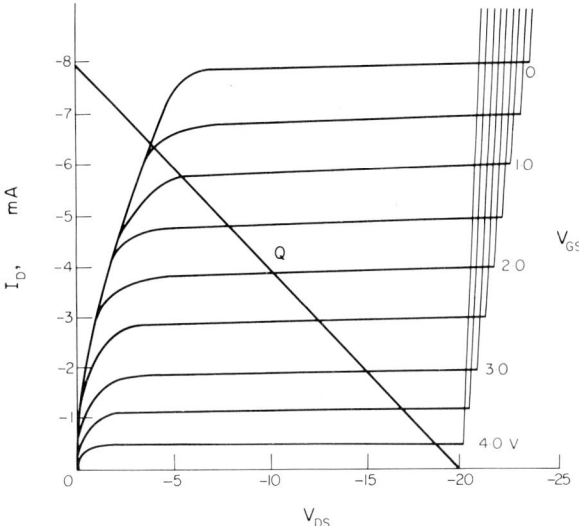

FIG. 1

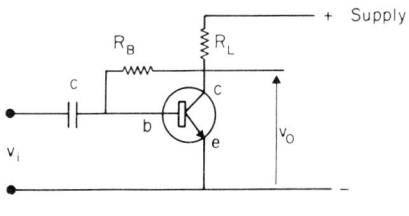

FIG. 2

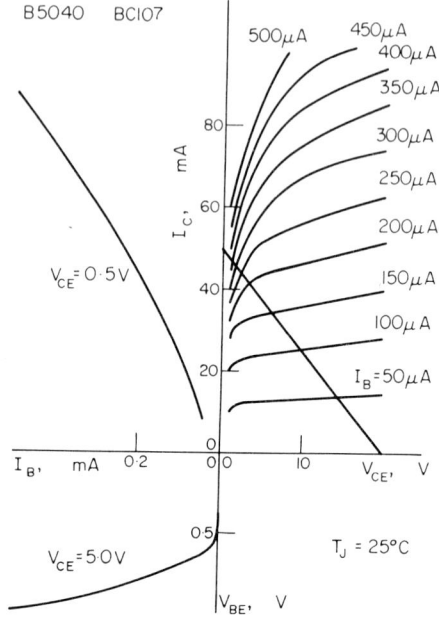

FIG. 3

CHAPTER 18

Amplifiers and Oscillators

WE shall now consider the application of some of the devices and components discussed in the previous chapters to the realisation of some simple amplifiers and oscillators. The technique adopted is to represent the active device chosen by its equivalent circuit, or model, and then to surround this with simple passive components arranged so as to achieve, as far as possible, the given requirements.

It will be clear from the details given previously about an equivalent circuit suitable for representing a pentode valve (p. 292) and that for representing a field effect transistor (p. 311) that these equivalent circuits are identical and only the symbols on the diagrams are different. The comparison between the symbols is as follows:

Pentode	F.E.T.
ρ	r_d
g_m	g_m
v_g	v_{gs}

It is, therefore, very easy to convert equations derived from circuits having pentode valves to those appropriate to similar circuits having f.e.t.s by simply changing the symbols.

It is not possible to derive such a comparison between the triode valve and the bipolar transistor because of the greater complication of the circuit model for the latter, however in many practical instances the representation of the active devices in a circuit can be reduced, without much error for small signal, low frequency analysis, to a simple current generator shunted by a resistor or a simple voltage generator in series with a resistor, and this has been done for most of the amplifier and oscillator circuits given in this chapter, which are shown some in valve and some in transistor form.

315

Resistance–Capacitance Amplifiers

R.C. amplifiers constitute the simplest useful embodiment of the amplifying valve and they are widely used in all types of communication equipment. A thorough understanding of their behaviour is therefore necessary.

In Ch. 16 we described in detail how the various d.c. connections were made to an R.C. coupled valve. Here we define the stage slightly differently so that each stage is identical with all the others.

In Fig. 18.1 we have included the grid resistor of V_2 in stage 1

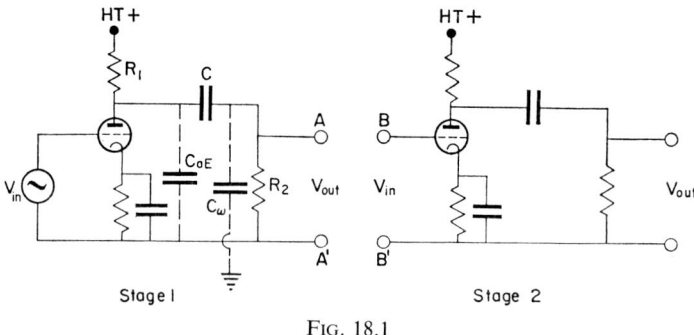

FIG. 18.1

because it forms part of the a.c. load of stage 1. Since the stages are identical, when AB and A^1B^1 are joined together, the overall gain is just the product of the individual gains.

The valves are drawn as triodes for simplicity, principles are the same for pentodes.

The operation of the R.C. amplifier is usually considered by dividing the gain versus frequency characteristic into three regions, the low-frequency region, the mid-band region and the high-frequency region. The measured frequency response is of the form shown in Fig. 18.2, with these regions indicated, and it can, in fact, be shown that all R.C. amplifier response curves have the same shape.

The drop in amplification in the low-frequency region is caused mainly by the reactance of C increasing to values which are comparable with the resistance value R_2, so that V_{out} is reduced below the a.c. voltage available at the anode terminal. The drop

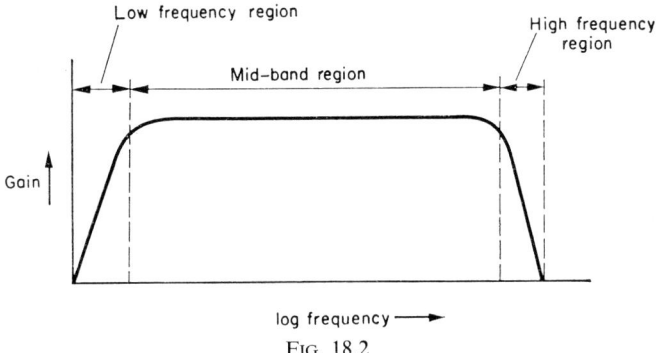

Fig. 18.2

may be caused partly by the cathode by-pass capacitor being inadequate but, with modern electrolytic capacitors, if this occurs, it is only due to intentional design.

In the mid-band region $1/\omega C \ll R_2$ and the gain is constant.

In the high-frequency region, the capacitance of the anode to ground and the capacitance between the anode wiring and ground plus the input capacitance of V_2, all provide reactance in parallel with R_2 which reduces the available output voltage.

We can now draw, for purposes of analysis, three equivalent circuits, valid for the three regions. We use the current generator equivalent circuit as it is, in general, simpler. The gain is A and R_{eff} is the combined value of the resistors.

Mid-band Region

From the equivalent circuit given in Fig. 18.3 we have:

$$A = \frac{V_{out}}{V_{in}} = -g_m R_{eff} \qquad (1)$$

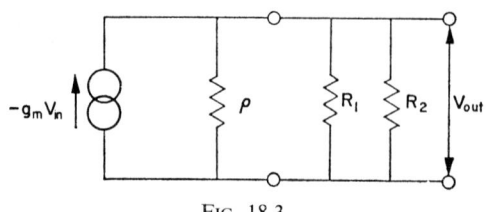

Fig. 18.3

where
$$\frac{1}{R_{\text{eff}}} = \frac{1}{\rho} + \frac{1}{R_1} + \frac{1}{R_2}$$

or
$$R_{\text{eff}} = \frac{\rho R_1 R_2}{R_1 R_2 + \rho(R_1 + R_2)}$$
(2)

Often, $R_1 \ll R_2$ and $R_1 \ll \rho$, in which case

$$A \simeq g_m R_1$$
(3)

Low-Frequency Region

Here the reactance of C is no longer negligible compared with R_2 and the equivalent circuit gives:

(a) *Approximate treatment* (Fig. 18.4). We assume that the reactance of C alters R_{eff} by only a negligible amount. Then

$$A = -g_m R_{\text{eff}} \times \frac{R_2}{R_2 - (j/\omega C)}$$

or
$$A = -g_m R_{\text{eff}} \frac{1 + (j X_2/R_2)}{1 + (X_2/R_2)^2}$$

where $X_2 = 1/\omega C$.

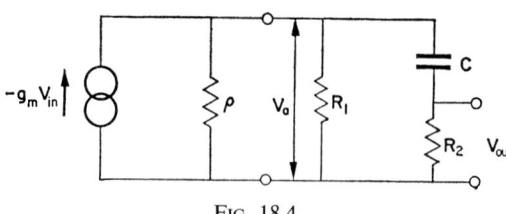

Fig. 18.4

We can write this as

$$A = A_{mid} \frac{1 + (jX_2/R_2)}{1 + (X_2/R_2)^2} \quad \text{where } A_{mid} = \text{mid-band gain}$$

$$|A| = \frac{A_{mid}}{\sqrt{[1 + (X_2/R_2)^2]}} = \frac{A_{mid}}{\sqrt{[1 + (1/\omega CR_2)^2]}} \tag{5}$$

$$\arg A = \tan^{-1}\left(\frac{X_2}{R_2}\right) = \tan^{-1}\left(\frac{1}{\omega CR_2}\right) \tag{6}$$

Let us now define a frequency ω_0, such that $\omega_0 CR_2 = 1$

i.e. $$R_2 = \frac{1}{\omega_0 C}$$

Then $$|A| = \frac{A_{mid}}{\sqrt{\left[1 + \left(\frac{\omega_0 CR_2}{\omega CR_2}\right)^2\right]}} = \frac{A_{mid}}{\sqrt{\left[1 + \left(\frac{\omega_0}{\omega}\right)^2\right]}} \tag{7}$$

$$\arg A = \tan^{-1}\left(\frac{\omega_0}{\omega}\right) \tag{8}$$

Then, when $\omega = \omega_0$ the voltage gain has been reduced to $1/\sqrt{2}\, A_{mid}$ or $0\cdot71\, A_{mid}$, and the power gain, which varies as A^2, has been halved or reduced by 3 dB, while the phase angle has reached the value 45°. In communication engineering the range of frequencies passed by any piece of apparatus is called the "bandwidth" and this quantity is usually measured as the frequency interval between the high frequency at which the power output drops to half and the low frequency at which the power drops to half. $F = \omega_0/2\pi$ is thus the low frequency bandwidth limit for an R.C. amplifier in cycles per second.

Also, notice that (7) and (8) with (14) and (15) below prove the earlier remark that the shape of R.C. response curves is always the same since $|A|$ and $\arg A$ are functions only of (ω_0/ω).

(b) *Exact treatment* (Fig. 18.5). The exact treatment follows the line of argument developed above. It is now simpler to use the voltage generator and Thévenin's theorem.

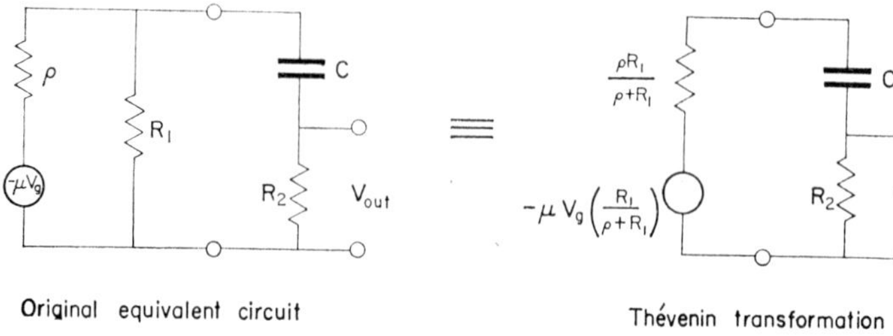

Original equivalent circuit   Thévenin transformation

FIG. 18.5

Using the Thévenin equivalent, Fig. 18.5, we first find the mid-band value when C is removed:

$$V_{\text{out}} = -\mu V_g \times \left(\frac{R_1}{\rho + R_1}\right) \times \frac{R_2}{\left(\frac{\rho R_1}{\rho + R_1}\right) + R_2}$$

$$A_{\text{mid}} = \left(\frac{-\mu R_1}{\rho + R_1}\right) \times \frac{R_2}{\left(\frac{\rho R_1}{\rho + R_1}\right) + R_2}$$

(Check that substitution $g_m = \mu/\rho$ converts this to $A_{\text{mid}} = -g_m R_{\text{eff}}$.)
Then, when C is important,

$$A = -\left(\frac{\mu R_1}{\rho + R_1}\right) \times \frac{R_2}{\left(\frac{\rho R_1}{\rho + R_1}\right) + R_2 - jX_2}$$

or $$A_{\text{mid}} = A \frac{\left(\frac{\rho R_1}{\rho + R_1}\right) + R_2 - jX_2}{\left(\frac{\rho R_1}{\rho + R_1}\right) + R_2}$$

$$= A \left[1 - j \frac{X_2}{\left(\frac{\rho R_1}{\rho + R_1}\right) + R_2}\right] \tag{9}$$

Comparison with the earlier results shows that we now have

$$\frac{X_2}{\left(\dfrac{\rho R_1}{\rho + R_1}\right) + R_2}$$

instead of X_2/R_2, so that our earlier result is valid if R_1 is much less than ρ and R_2. The exact expression for ω_0 is

$$\omega_0 C \left[\left(\frac{\rho R_1}{\rho + R_1}\right) + R_2 \right] = 1 \tag{10}$$

High-Frequency Region

Here the equivalent circuit is as shown in Fig. 18.6 where

$$C_2 = C_{ac} + C_\omega + C_{gc}$$

(where C_{ac} is capacitance from anode to cathode, C_{gc} from grid to cathode and C_ω from connections to earth)

$$\simeq 25 \text{ pF for most modern valves.}$$

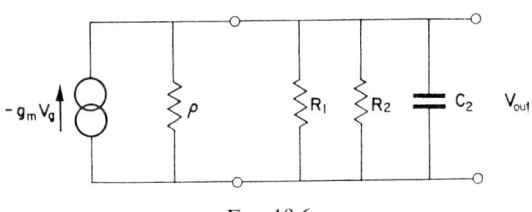

FIG. 18.6

The total admittance is $G + jB$, where

$$G = \frac{1}{\rho} + \frac{1}{R_1} + \frac{1}{R_2} = \frac{1}{R_{\text{eff}}} \qquad B = \omega C_2$$

$$\therefore \ V_{\text{out}} = \frac{-g_m V_g}{G + jB} \quad \text{or} \quad A = \frac{-g_m}{\dfrac{1}{R_{\text{eff}}} + j\omega C_2} = \frac{g_m R_{\text{eff}}}{1 + j\omega C_2 R_{\text{eff}}}$$

Therefore $\qquad A = \dfrac{A_{\text{mid}}}{1 + j\omega C_2 R_{\text{eff}}}$ $\qquad$ (11)

$$|A| = A_{\text{mid}} \dfrac{1}{\sqrt{[1 + (\omega C_2 R_{\text{eff}})^2]}}$$ $\qquad$ (12)

$$\arg A = \tan^{-1}(-\omega C_2 R_{\text{eff}})$$ $\qquad$ (13)

Then if we now define ω, by putting $\omega_1 C_2 R_{\text{eff}} = 1$,

$$|A| = A_{\text{mid}} \dfrac{1}{\sqrt{[1 + (\omega/\omega_1)^2]}}$$ $\qquad$ (14)

$$\arg A = \tan^{-1}(-\omega/\omega_1)$$ $\qquad$ (15)

Again, the voltage gain is reduced to $1/\sqrt{2}$ when $\omega = \omega_1$ and ω_1 represents the higher bandwidth point. The phase lags by $45°$ at $\omega = \omega_1$ instead of leading. As before, $|A|$ and $\arg A$ are functions only of (ω/ω_1).

The bandwidth of the amplifier is thus given by

$$\frac{\omega_1 - \omega_0}{2\pi} = \frac{1}{2\pi}\left(\frac{3}{C_2 R_{\text{eff}}} - \frac{1}{CR_2}\right) \text{c/s}$$

Notice that the conditions for high-frequency and low-frequency response are independent of one another and so are entirely at the disposal of the designer.

A common problem encountered in practice is that of achieving a desired value of gain over a very wide frequency band. Let us consider what happens if we wish to obtain gain A over a bandwidth of x megacycles. C_2 is more or less fixed by the valves, so that R_{eff} has to be fixed by

$$R_{\text{eff}} = \frac{10^6}{2\pi x C_2}, \quad C_2 \text{ in picofarads, } x \text{ in Mc}$$

For $C_2 = 25$ and $x > 1$, R_{eff} cannot exceed $7\,\text{k}\Omega$ and we can certainly under such circumstances approximate R_{eff} by R_1.

Then $\qquad R_1 \simeq \dfrac{7}{x}\,\text{k}\Omega \quad$ and $\quad A_{\text{mid}} = \dfrac{-g_m 7}{x}$

For valves used in this service g_m lies between 5 and 12 mA/V

$$\therefore A_{\text{mid}} \text{ lies between } 35/x \text{ and } 84/x$$

In a T.V. receiver $x \simeq 5$ Mc and in broadband telecommunication systems bandwidths of 20 Mc/s are required, so only small stage gains can be achieved.

This illustrates one of the very important principles of communication engineering, that in any piece of apparatus the product of gain and bandwidth tends to be constant and one can only improve the performance by improving the active circuit elements, in this case the valves. Of course, by bad design it is always possible to make less than optimum use of the active elements.

R.C. Stages in Cascade

We have already stated that the overall gain is the product of the individual gains. This automatically means that the bandwidth will be decreased. For instance at the 0·71 voltage points, or half-power points, for the single stage a two-stage amplifier gives a gain of half. If the stages are identical and there are n in cascade then for the low frequency end we have

$$|A|^n = (A_{\text{mid}})^n \left(\frac{1}{\sqrt{[1 + (\omega_0/\omega)]^2}} \right)^n$$

Usually we require to know the half-power bandwidth point for n stages. The required value is given by putting

$$\sqrt{[1 + (\omega_0/\omega)^2]} = 2 \times 1/n$$

and solving for ω.

The high frequency half-power point is found in a similar way by solving

$$\sqrt{[1 + (\omega_1/\omega)^2]} = 2 \times 1/n.$$

Confusion is sometimes caused by the terms voltage, current and power amplifiers. In general, any amplifier must give a power gain and will amplify either voltage or current or both.

However, amplifiers can be adapted to give optimum performance as either voltage, current or power amplifiers, by choosing
the input resistance. Thus, an amplifier will give optimum voltage
gain when the connection of the amplifier to the generator, of
internal impedance R_g, does not much alter the open circuit
generator voltage, thus the input impedance of the amplifier,
Z_{in}, should obey the condition $Z_{in} \gg R_g$. For optimum current
performance, $Z_{in} \ll R_g$ so that almost the short circuit current
will flow.

For optimum power gain, Z_{in} should be adjusted until the
maximum power is extracted from the generator, or in other
words, the amplifier input circuit is matched to the generator.

Feedback

If a portion of the output from an amplifier is connected back
to the input, we say that *feedback* has been applied. If the feedback part of the output voltage is in phase with the input, the
output will increase, and we speak of positive feedback, while if
the feedback portion is 180° out of phase with the input the output
and effective gain are reduced and the feedback is negative.

We shall find that negative feedback gives important advantages in freedom from distortion, freedom from changes in
overall gain due to the changes in valve and component characteristics which occur with life, etc., which far outweigh the disadvantage of reduced gain. In particular, all amplifiers used in
electronic measuring instruments should be designed with negative feedback for a gain characteristic which is constant in time.

Figure 18.7 shows, in block diagram form, an amplifier with
feedback.

The amplifier has gain A so that without feedback

$$V_o = A \, V_i \qquad (16)$$

A is complex because the amplifier has phase shift as well as
gain. A portion of the output voltage V_o is fedback to the input.
This is

$$\beta \, V_o$$

where β is complex, to allow for phase shift in the coupling network.

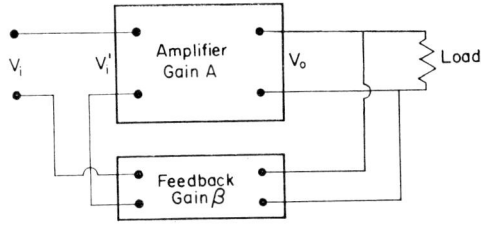

FIG. 18.7

The actual input voltage at the terminals of the amplifier is now

$$V_i^1 = V_i + \beta V_o \qquad (17)$$

$$V_o = AV_i^1 = AV_i + A\beta V_o$$

or
$$\frac{V_o}{V_i} = \frac{A}{1 - A\beta} \qquad (18)$$

Thus, the feedback has changed the gain from A to $A/1 - A\beta$. Whether this is a decrease or an increase depends on $|1 - A\beta|$. If this is < 1 the feedback is positive and the amplifier is regenerative. If $|1 - A\beta| \rightarrow 0$, the gain tends to $\rightarrow \infty$ and the amplifier oscillates while if $|1 - A\beta| > 1$, the feedback is negative and the gain is reduced. $A\beta$ is called the *feedback factor* and it is usually complex, so we can write

$$A\beta = a + jb \qquad (19)$$

Negative Feedback Amplifiers

We have just defined the condition of negative feedback (or N.F.B.) by saying $|(1 - a) - jb| > 1$

or
$$[(1 - a)^2 + b^2] > 1 \qquad (20)$$

This can be satisfied by a either large and negative or large and positive. When a is large and positive there will usually be some different frequency which will make $a \simeq 1$ and the amplifier would be unstable and would tend to oscillate at that frequency. This state of affairs cannot happen when a is negative and we are usually interested in this condition, i.e. when the feedback voltage reduces the input voltage but is not necessarily 180° out of phase.

Case (i) $|1 - A\beta| \gg 1.$

For a negative and large $1 - \overrightarrow{A\beta} \simeq - \overrightarrow{A\beta}$

$$\therefore \frac{V_o}{V_i} = \frac{\overrightarrow{A\beta}}{1 - \overrightarrow{A\beta}} = \frac{\overrightarrow{A}}{-\overrightarrow{A\beta}} = \frac{-1}{\overrightarrow{\beta}}$$

Thus, the negative feedback has changed, and in fact reduced, the amplifier gain to $-1/\beta$. But the gain is now independent of A and depends only on β. A is an active network containing valves or transistors which change their characteristics with age, operating voltages, etc. β depends only on the feedback network and this can be purely passive, containing perhaps only resistances and capacitors. Thus, provided always that A remains large enough to make $A\beta \gg 1$, the feedback network will determine the effective gain.

Example: Suppose that a certain equipment uses valves in stages with gains of 100 and the valves are regularly tested and are replaced when g_m is reduced to 50 per cent of the value when new. Then the gain will have dropped to 50.

Two such stages with 1 per cent negative feedback will give a gain of approximately 100. Then during life the gain will vary from

$$\frac{10,000}{101} = \frac{100}{101} = 99 \quad \text{to} \quad \frac{2500}{26} = \frac{100}{104} = 96 \qquad \text{or by 3 per cent.}$$

Alternatively g_m could drop to 10 per cent of the initial value before the gain would drop to half. Either the performance is much improved or much longer valve life is obtained for the same performance.

Apart from the constant gain, there are other advantages. These include:

(i) Reduction of *harmonic distortion*. Since valve characteristics are not, in fact, straight lines all amplifiers are non-linear to some extent and this gives rise to harmonic distortion, i.e. to the presence of frequencies in the output waveform which are not present in the input.

When, by use of N.F.B., the amplifier gain only depends on the β network, which contains only passive, linear circuit elements this source of distortion is eliminated.

In the case when A and β are real (no phase shifts), it is easy to show that the harmonic distortion at a given output voltage is reduced by the factor $(1 - A\beta)$. When A and β are complex a simple expression cannot be obtained.

(ii) Control of *frequency response*. When the N.F.B. is great enough to make the gain equal to $-1/\beta$, then the frequency response will be closely equal to the frequency response of the network and independent of the frequency response of the amplifier.

Thus, if β is derived from a purely resistive network then the gain will be constant until the frequency becomes so high that unity can no longer be neglected in comparison with $A\beta$. It is simpler to design a passive network with the required frequency characteristics than it is an active network.

(iii) Reduction of *output impedance*. In Fig. 18.8(a) without feedback, the amplifier has gain A output impedance Z_o. With (large) N.F.B. and infinite load impedance, i.e. no load, we have

$$V_o = A(V_i + \beta V_o)$$

$$V_o(1 - \beta A) = A V_i \tag{21}$$

When Z_L is connected, (b), the current is i and $V_o^1 = Z_L$ $\quad$ (22)

We must remember that the new open-circuit output voltage is now modified to $A(V_i + \beta V_o^1)$ so $V_o^1 = A(V_i + \beta V_o^1) - iZ_o$.

(a)

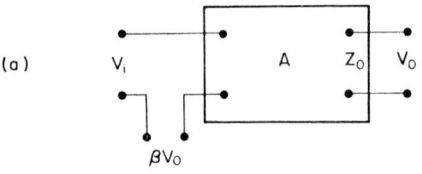

Before connecting load

(b)

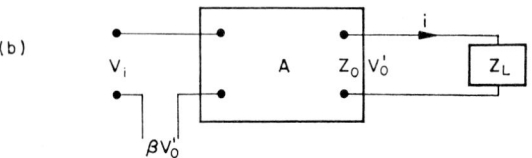

After connecting load

FIG. 18.8

Using (22) $V_o^1 = A(V_i + \beta V_o^1) - V_o^1 \dfrac{Z_o}{Z_L}$

or $V_o^1 \left(1 - A\beta + \dfrac{Z_o}{Z_1}\right) = A V_i$ (23)

But, from (21) we can replace $A V_i$

$\therefore \; V_o^1 \left(1 - A\beta + \dfrac{Z_o}{Z_L}\right) = V_o(1 - A\beta)$

$\therefore \dfrac{V_o^1}{V_c} = \dfrac{1 - A\beta}{1 - A\beta + (Z_o/Z_L)} = \dfrac{Z_L}{Z_L + [Z_o/(1 - \beta A)]}$ (24)

This equation tells us how the load voltage varies as a function of Z_L, the load impedance, and we can therefore draw a Thévenin equivalent circuit for the output. Clearly, the output impedance has been reduced from Z_o to $Z_o/(1 - \beta A)$ by the action of N.F.B. Note that *series* N.F.B. is necessary to do this.

To understand the physical significance of these results we can confine ourselves to the simple case $A\beta$ real, $|A\beta| \gg 1$.

In this condition, the input voltage V_i is almost entirely cancelled by the feedback voltage βV_o and it is the difference between the two which is applied to the input terminals and amplified. This difference is a small percentage of V_i so that any change in V_o, and therefore in βV_o, will cause a much larger fractional change in the voltage at the amplifier terminals. However if V_o is not a faithful replica of V_i then the difference between V_i and βV_o will be increased, and, since it is applied in the negative direction, will tend to reduce the output. This all amounts to saying that the gain now only depends on $-1/\beta$ and therefore not on A.

We can understand the change of output impedance in the same way. When the amplifier is loaded the output voltage and therefore $|\beta V_o|$ decrease, so that more effective input is applied to the amplifier and the output rises in such a manner as to maintain the difference between V_i and βV_o constant. Thus, V_o is much more nearly dependent of load than it is in an amplifier without feedback. By definition such an amplifier has a low output impedance.

Note that it is impossible for the feedback signal to exactly cancel the input. This would require infinite amplifier gain. However, the larger A is made the more nearly the signals cancel and we next enquire into the limitations on increasing A.

Since βA is, over a large frequency band, complex, one of the most useful methods for presenting information is to prepare a polar plot of βA over as wide a frequency range as possible. These plots are called Nyquist diagrams and an example is given in Fig. 18.9.

Note, that the points for equal frequency increments may cluster closely together in some regions and be far apart in others. For example, in the mid-band region of an R.C. amplifier, A is constant, and if β is derived from a resistance network, β is constant. Thus the whole mid-band region condenses to a single point.

Let I be the point $(1, 0)$ on the diagram. OF is the vector representing βA at a frequency ω. By vector subtraction FI is the vector $(1-\beta A)$. We have already seen that the amplifier is regenerative when $|(1-\beta A)| < 1$ and degenerative when $|(1-\beta A)| > 1$. Therefore if we draw a unit circle with I as centre, the amplifier will be regenerative over the range of frequencies for which the Nyquist plot lies inside the unit circle.

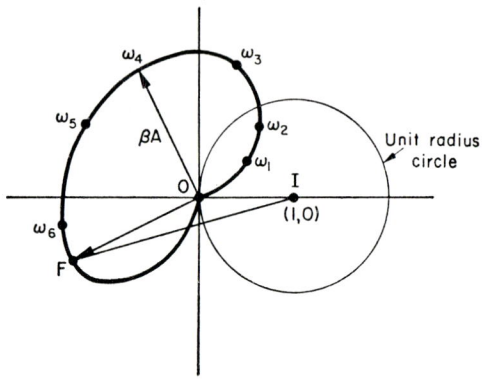

Fig. 18.9

We have already seen that, if the plot goes through I, $|(1-\beta A| = 0$ and the gain is infinite, and the amplifier will oscillate at the corresponding frequency. Nyquist has shown that a more general condition is that the amplifier is unstable if the plot encloses $(I, 0)$.

Suppose now that we increase $|A|$ without change of phase, then if the original Nyquist plot is of such a shape that it encloses I, when all the vectors are multiplied by the scale factor the amplifier will become unstable. If this is the case, there is obviously a limiting value of gain which must not be exceeded.

The discussion of the conditions which lead to stable and unstable cases is difficult and cannot be discussed here.

Voltage and Current Feedback

In all the above we have tacitly assumed voltage feedback, i.e.

$$V_o = \phi(V_i)$$

Current feedback, where $i_o = \phi(V_i)$, is also possible.

In Fig. 18.10(a) β depends on a potentiometer which is independent of frequency up to very high frequencies when phase-shifts become important.

In Fig. 18.10(b) the feedback voltage depends on the current through the load and on the characteristics of Z_1.

(a)

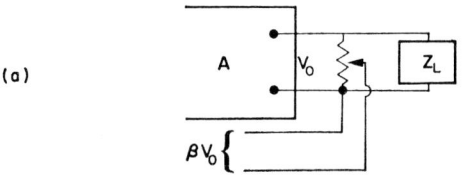

(b)

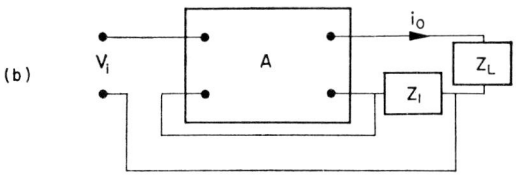

FIG. 18.10

We can deduce the general properties of current feedback by physical reasoning, if we consider the case when $A\beta \gg 1$, i.e. when $i_o Z_1 \simeq V_i$.

Since the feedback voltage depends on the output current the feedback acts so as to make the ratio i_o/V_i independent of the characteristics of A, i.e. parameters of the active devices.

The frequency variation will depend almost entirely on the frequency variation of Z_1.

The output current i_o will tend to be an accurate replica of the input voltage V_i.

For example in magnetic deflection systems for T.V. the deflection current we require is a saw-tooth waveform in an inductive load. We can easily generate a saw-tooth voltage and can then achieve the saw-tooth current by current feedback methods.

If the load is varied the current tends to be constant, i.e. the amplifier is a constant current source and the N.F.B. has increased greatly the internal impedance Z_o.

The input impedance of any amplifier is finite although the real part may be very large. As the frequency increases the inevitable capacitance across the input terminals becomes the major component and may easily represent an undesirably low reactance.

When N.F.B. is applied in series with the input, as we have done, the voltage across the amplifier terminals is greatly reduced and so is the current flowing from the generator. The input impedance is much increased.

There are other ways of applying feedback to the input and it is also possible to reduce the input impedance.

Practical Design of Feedback Circuits

The design of other than simple N.F.B. amplifiers is complicated by the difficulty of satisfying the Nyquist stability criterion. This involves study of the amplifier gain at all frequencies as instability may result at frequencies entirely outside the normal pass-band. At high frequencies stray capacitances are important and at low frequencies phase-shifts due to coupling and by-pass capacitors are not negligible. The phase-shift of 180° from each valve grid to anode is naturally very important. Thus, for n valves in cascade we have a factor of

$$- 1 \text{ when } n \text{ is odd}$$
$$+ 1 \text{ when } n \text{ is even}$$

In particular for n odd a purely resistive network across the output will give N.F.B.

The two following examples show N.F.B. incorporated in simple, single transistor circuits.

In Fig. 18.11(a). C_1 serves as an isolator and can be ignored except at low frequencies. $\beta = R_2/(R_1 + R_2)$ and $R_1 + R_2$ should be large. This is voltage N.F.B.

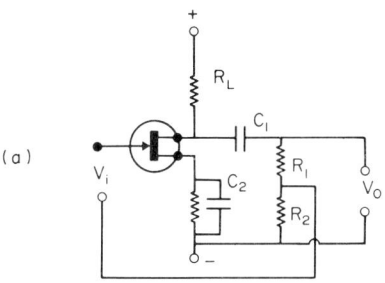

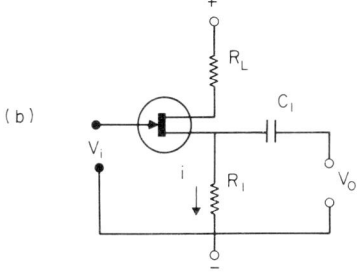

FIG. 18.11

In Fig. 18.11(b) the net input voltage

$$V_g = V_i - iR_1$$

so that current feedback is in operation. Thus, by merely leaving out the by-pass capacitor, C_2, in (a) we gain the advantages of N.F.B. but lose some gain.

The Cathode Follower

The circuit of Fig. 18.12(a) is a special case of Fig. 18.11(b), applied to a valve, in which there is no series resistance in the anode lead and a relatively large ($\simeq 2\rho$) value of resistance in the

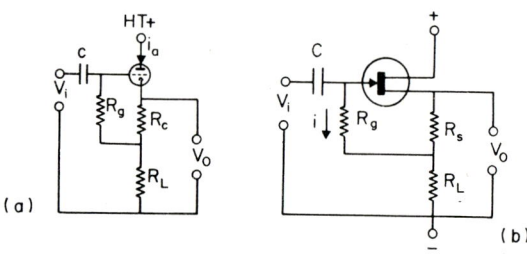

FIG. 18.12

cathode lead. The load line is constructed using E and R_L in the ordinary way. The current is then found for a given grid bias and R_c can be chosen so that $i_a R_c = -V_g$. R_c will be $\ll R_L$ so that its presence will not modify the load line. The grid resistor is then connected as shown to derive the bias. The voltage across R_L, $i_a R_L$, is now large and the current feedback will operate for input voltages V_i up to $i_a R_L$. The reactance of $C \ll R_L$.

The voltage feedback is

$$V_o \times \frac{R_L}{R_L + R_C} \simeq V_o$$

since $R_L \gg R_c$. Thus $\beta \simeq 1$ and the input impedance is very high, the output impedance very low and V_o very closely reproduces V_i. The approximate direct analysis is as follows:

Neglecting R_c $\qquad\qquad V_g \simeq V_i - V_o$ $\qquad\qquad$ (25)

$$i_a = \frac{1}{\rho}\,(-V_o + \mu V_g) = \frac{1}{\rho}\,(-V_o + \mu V_i - \mu V_o)$$

But $\qquad\qquad V_o = R_L i_a = \frac{R_L}{\rho}\,[\mu V_i - (1+\mu)V_o]$ $\qquad\qquad$ (26)

or $\qquad\qquad V_o\left[1 + \frac{R_L}{\rho}\,(1+\mu)\right] = \frac{R_L \mu V_i}{\rho}$

$$\frac{V_o}{V_i} = \frac{\mu R_L}{\rho + R_L + \mu R_L}$$ $\qquad\qquad$ (27)

The gain is < 1, but μR_L is much greater than R_L and also than ρ so gain $\rightarrow 1$.

The output impedance is very easily found by finding the ratio of the open-circuit output voltage to the short-circuited output current. From (27)

$$V_o = \frac{\mu R_L V_i}{\rho + R_L + \mu R_L}$$ $\qquad\qquad$ (28)

The short-circuit output current is found by observing that when the output is short circuit the feedback voltage is removed and the full value V_i is applied to the grid.

$$\therefore i_{sc} = g_m V_i$$ $\qquad\qquad$ (29)

and $\qquad\qquad Z_o = \frac{V_o}{i_{sc}} = \frac{1}{g_m}\,\frac{\mu R_L}{\rho + R_L + \mu R_L} \simeq 1/g_m$ $\qquad\qquad$ (30)

The approximation is better the nearer the gain approaches unity. If $g_m = 12\cdot5$ mA/V

$$Z = \frac{1000}{12\cdot5} = 80\ \Omega$$

Cathode followers are therefore used to couple high impedance devices to coaxial cables with characteristic impedances between 50 and 100 Ω.

Since the cathode follower has a large input impedance and a low output impedance and a voltage gain of near unity, it gives a very large power gain. It may be characterized as an impedance transforming device which operates over a much larger frequency range than does a transformer.

The name cathode follower derives from the fact that the potential of the cathode relative to ground is very nearly equal to V_i so that the cathode "follows" the grid. Note that V_o is in phase with V_i.

One of the simplest types of servo system is that used for remote indication of the reading of some instrument. This is done by amplifying the difference between the indication of the remote instrument and the original. The close nature of the analogy between this and negative feedback will at once be appreciated.

The Source Follower

Fig. 18.12(b) shows the same circuit applied to an f.e.t. when the arrangement is then called a *source follower*.

The working is as follows, assuming a current generator as the equivalent circuit for the f.e.t.

$$v_i - \left(\frac{R_L}{R_s + R_L}\right) v_o \simeq iR_g$$

$$\frac{v_o}{v_i} \simeq \frac{g_m(R_s + R_L)}{1 + g_m(R_s + R_L)}$$

$$v_i\left(1 - \frac{R_L g_m}{1 + g_m(R_s + R_L)}\right) = iR_g$$

$$\therefore \text{ Input resistance } R_i = \frac{v_i}{i} = \frac{R_g \left[1 + g_m (R_s + R_L)\right]}{1 + g_m (R_s + R_L) - g_m R_L}$$

$$= \frac{R_g \left[1 + g_m (R_s + R_L)\right]}{1 + g_m R_3}$$

but $R_s \ll R_L$

$$\therefore R_i \simeq R_g \left[1 + g_m (R_s + R_L)\right]$$

For typical values this input resistance would be as high as about 50 MΩ and the main advantage of the circuit is that it gives a very high input impedance and a low output impedance. It is thus almost invariably used to link various types of electrical transducer, which typically are unable to give any appreciable current output, to the remainder of an appropriate amplifying system.

Low-power Oscillators

Low-power oscillators are used for measurement purposes, as local oscillators in superheterodyne receivers and as sources for the control of pulse generators, etc. By low power we mean oscillators such that the valves operate in a fairly linear part of their characteristics. The high-power oscillators used in transmission equipment usually operate in a high-efficiency mode which we shall discuss later. The general principles of operation are as follows:

(a) an amplifier works as a convertor of d c. energy, derived from the batteries, to high frequency energy.

(b) the output power is very much greater than the input power, thus by returning a small proportion of the output power to the input it should be possible to maintain a steady output even when any external source is removed. A device in which this is done is an oscillator.

In the simple arrangement shown in Fig. 18.13 the coupling network forms the load on the amplifier and returns (feeds back) a certain percentage of the output voltage to the input:

Let V_o = output and V_i = input, then

$$V_o = A\,\mathrm{e}^{\mathrm{j}\theta} . V_i \qquad (31)$$

(i.e. using $\mathrm{e}^{\mathrm{j}\theta} = \cos\theta + \mathrm{j}\sin\theta$).

A = voltage gain, θ = phase shift.

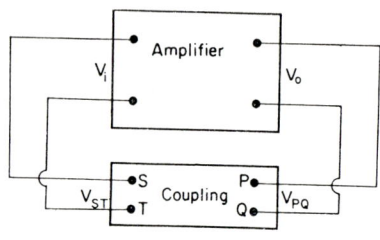

FIG. 18.13

Similarly for the coupling network

$$\frac{V_{ST}}{V_{PQ}} = B\,\mathrm{e}^{\mathrm{j}\phi} = \frac{V_i}{V_o} \qquad (32)$$

where

$$B = \left|\frac{V_{ST}}{V_{PQ}}\right|,$$

and $\qquad \phi$ = phase shift of V_{ST} relative to V_{PQ}.

If V_i is to be steadily maintained, we must have

$$V_i = V_i\,A\,\mathrm{e}^{\mathrm{j}\theta} . B\,\mathrm{e}^{\mathrm{j}\phi}$$

or $\qquad\qquad AB\,\mathrm{e}^{\mathrm{j}(\theta+\phi)} = 1 \qquad (33)$

This leads to the conditions

$$\left. \begin{array}{l} AB = 1 \\ \theta + \phi = 0,\ 2n\pi, \qquad n \text{ an integer} \end{array} \right\} \qquad (34)$$

A, B, θ and ϕ are all functions of frequency and there will only be one or, at most, a few frequencies for which both conditions are simultaneously obeyed. Normally, operation is achieved by varying A.

As a rule, we wish to design the amplifier to work at a pre-determined frequency and a useful way of doing this is to make either θ or ϕ vary very rapidly with f in the region of oscillation. In particular a parallel tuned circuit has the property that the phase, relative to the input, of the voltage across the circuit changes very rapidly by almost $180°$ in going through resonance, if the Q is high. This will ensure that the phase condition can only be satisfied near the resonant point of the circuit.

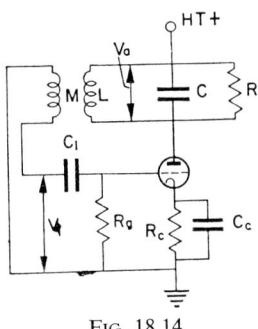

FIG. 18.14

The tuned anode oscillator having a resonant circuit in the anode lead is shown in Fig. 18.14. The analysis which follows is made on the following assumptions:

1. Valve operation is linear.

2. R_g, C_1 and C_c all so large that they do not influence the r.f. performance.

3. R includes loss in L and C effects due to ρ and any added load resistance.

Then
$$i = g_m V_i = V_a \left(\frac{-j}{\omega L} + j\omega C + \frac{1}{R} \right) \qquad (35)$$

$$V_i = \pm j\omega M \frac{V_a}{j\omega L} = \pm \frac{M}{L} V_a \qquad (36)$$

Divide equation (35) by equation (36)

$$g_m = \pm \frac{L}{M} \left(j\omega C - \frac{j}{\omega L} + \frac{1}{R} \right) \tag{37}$$

Equating real and imaginary parts

$$g_m = \pm \frac{L}{RM} \tag{38}$$

$$j\left(\omega C - \frac{1}{\omega L} \right) = 0 \quad \text{or} \quad \omega = \frac{1}{\sqrt{(LC)}} \tag{39}$$

Thus, the circuit will oscillate at the self-resonant frequency of the circuit when g_m is adjusted (e.g. by varying the grid bias) so that equation (38) is fulfilled.

This analysis is a steady state analysis: it tells us what happens when the oscillator is running but it cannot tell us anything about the way in which the oscillations build up or what would happen if, for example, $g_m > L/RM$. To study the transient conditions we have to write down the differential equations which are valid for V_i, V_a and i of any arbitrary time dependence and not merely sinusoids.

We have $\quad i = g_m V_i = \dfrac{V_a}{R} + C \dfrac{dV_a}{dt} + \dfrac{1}{L} \int V_a dt$

$$i_L = \frac{1}{L} \int V_a dt$$

$$\therefore \; V_i = \pm \frac{M}{L} \frac{d}{dt} \left(\int V_a dt \right) = \pm \frac{M}{L} V_a$$

or $\quad \dfrac{V_a}{R} + C \dfrac{dV_a}{dt} + \dfrac{1}{L} \int V_a dt \pm g_m \dfrac{M}{L} V_a = 0 \tag{40}$

Differentiate w.r.t. t

$$\frac{d^2 V_a}{dt^2} + \left(\frac{1}{RC} \pm \frac{g_m M}{LC} \right) \frac{dV_a}{dt} + \frac{V_a}{LC} = 0 \tag{41}$$

V_a has been chosen to minimize algebra. We could have solved for the loop currents in which case the final result would have been of the same form as (41).

Equation (41) is an ordinary second order differential equation with constant coefficients.

Put $\qquad \left(\dfrac{1}{RC} \pm \dfrac{g_m M}{LC}\right) = 2k, \qquad \dfrac{1}{LC} = \omega^2$

Case (i): $k < \omega$

Solution by standard means is

$$V_a = e^{-kt} A \cos[\sqrt{(\omega^2 - k^2)}t + \phi] \qquad (42)$$

The constants A and ϕ will be found from the initial conditions.

Since the magnitude and sign of M are under our control k may be positive, negative or zero.

k positive. Damped oscillations, which eventually die out.
$k = 0$. Stable oscillations of constant amplitude.
k negative. Oscillations which increase exponentially.

The change from one condition to the next is continuous.

1. With k positive the anode circuit behaves as an ordinary resonant circuit with effective resistance depending on k, as k decreases the parallel resistance and Q increase. A signal fed into the grid from an outside source eventually decays.

2. $k = 0$. In this special case Q is infinite. A coupled signal will just be maintained.

The conditions are

$$\frac{1}{R} = \frac{g_m M}{L}$$

$$\sqrt{(\omega^2 - k^2)} \to \omega = \frac{1}{\sqrt{(LC)}}$$

These are identical with the steady state condition.

3. k negative, i.e. $g_m M/L > 1/R$. (Amplitude limitation.)

In practice it will not be possible to maintain oscillation with

$k = 0$; for instance, if the oscillation were temporarily inter-rupted it might not restart and so k has to be very slightly negative. This does not, however, mean that the oscillation builds up according to e^{kt} to infinite amplitude. The actual amplitude reached will be finite because as the oscillation in-creases in amplitude the average value of g_m diminishes because of the non-linearity of the valve characteristics. Also, for big amplitudes grid current may flow which loads the input circuit.

If, for very small amplitudes, k is negative then the very small thermal noise voltages in the anode circuit, which have com-ponents at the resonant angular frequency $1/\sqrt{(LC)}$, will be amplified and the oscillation will build up. As it builds up k gets less negative and the rate of growth diminishes. Eventually stable operation results around the valve for which $k = 0$.

It is easy to see that when the oscillation is stable, the power absorbed in the grid cathode circuit must equal the power fed back from the anode. But a detailed calculation on purely theoretical bases is difficult for the following reason: in general instead of $i = g_m V_g$ we shall have

$$i = g_m V_g - b V_g^2 - C V^3, \text{ etc.}$$

and if we substitute this expression for i in our differential equation we find that the result is a non-linear differential equation. Such equations can only, at present, be solved by approximation methods, although much research is done in this field of mathematics. We do know, however, that such equations do predict stable oscillator amplitudes whereas linear equations can tell us nothing whatsoever about the amplitude.

When k is negative, the oscillation frequency will be a little less than $1/2\pi \sqrt{(LC)}$. However, from the arguments already given, when stable conditions are reached k will not be much different from zero. Thus, $\omega \simeq 1/\sqrt{(LC)}$, as in a linear system. This holds with considerable accuracy by ordinary engineering standards but is not sufficiently good for precision measurements.

Apart from the physical understanding we have gained, the differential equation (41) tells us:

(i) the condition for build-up, $g_m M/L > 1/R$

(ii) the oscillation frequency, $\omega^2 = 1/LC$ $\quad \therefore \ \omega = 1/\sqrt{(LC)}$

(iii) the rate of build-up, $k = \frac{1}{2}\left(\dfrac{1}{RC} \pm \dfrac{g_m M}{LC}\right)$

The first two are identical with the results of the steady-state analysis and the third is of little practical importance. Thus the j-operator method would give us the results we use but the differential equation is a much more powerful method and can be used for much more complicated problems. Here, we are using it on a very simple example.

Typical Examination Questions

1. The following questions relate to the resistance-coupled amplifier shown in Fig. 1. The triode A has an amplification factor of 30 and a slope resistance of 20 kΩ, and is biased by the resistance to a substantially linear portion of the characteristics. The reactance of C_1 and C_3 are negligible at all relevant frequencies. V_1 is the voltage applied to the input terminals of the first stage and V_2 is the voltage which it produces at the input terminals of the second stage.

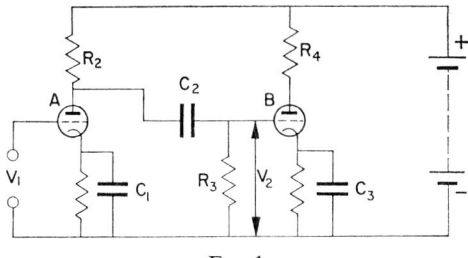

FIG. 1

(a) If the reactance of C_2 is negligible and $R_3 = 100$ kΩ what is the maximum value of the ratio V_2/V_1 which can be obtained with any value of R_2?

Ans: 25

(b) What must be the value of R_2 if the ratio V_2/V_1 is to be 80 per cent of the maximum value found in (a), if C_2 and R_3 are the same as before?

Ans: 66·7 kΩ

(c) If $R_2 = R_3 = 100$ kΩ and $C_2 = 0.01$ μF, calculate the value of $|V_2/V_1|$ and the phase angle of V_2 relative to V_1 for frequencies of 4 kc/s and 20 c/s respectively.

Ans: 21·4, leading $(\pi + \tan^{-1} 0.034)$; 3·1, leading $(\pi + \tan^{-1} 4.8)$

(d) With $R_2 = 100$ kΩ, $R_3 = 400$ kΩ and $C_2 = 0.1$ μF, stray capacities are found to be equivalent to condensers of 15 μμF in parallel with R_2 and 10 μμF in parallel with R_3 respectively. What is the value of $|V_2/V_1|$ at a frequency of 1 Mc/s and what is the phase angle of V_2 relative to V_1?

Ans: 8·9, lagging $(\pi + \tan^{-1} 2.5)$

(e) The mutual conductance of V_2 is 3 mA/V and its amplification factor is 6. The load resistance R in its anode circuit can be varied. What is the maximum a.c. power that can be dissipated in R_4 when $V_2 = 1$ V r.m.s.?

Ans: 4·5 mW

2. An amplifying stage has the circuit of Fig. 2. The mutual conductance of the pentode under the conditions of operation is 1 mA/V and the anode slope resistance, which should be taken into account, is 1 MΩ. The valve is to be used with a screen potential of 140 V and a grid bias of -2 V, and there is a resistive load of 1 MΩ connected as shown. The stray capacitance $C = 20$ pF, $R_3 = 200$ kΩ, and $R_1 = 1$ MΩ. The mean anode current is 1 mA and the mean screen current 0·25 mA.

(a) Calculate the values of R_2 and R_4.

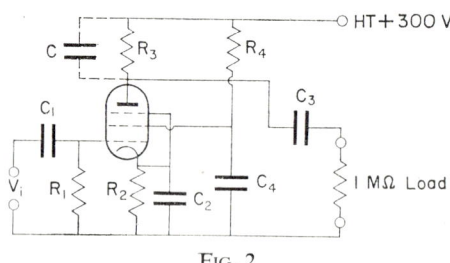

Fig. 2

Ans: $R_4 = 632$ kΩ, $R_2 = 1600$ Ω

(b) Suggest minimum values for C_1, C_2, C_3 and C_4 if the amplifier is to operate at frequencies down to 10 c/s.

Ans: $C_1 = 0.2$ μF, $C_2 = 200$ μF, $C_3 = 0.2$ μF, $C_4 = 0.5$ μF
(These values are not critical.)

(c) What is the mid-frequency gain and at what frequency will the presence of C have caused this gain to fall by 20 per cent? What will then be the phase of v_a with respect to v_g?

Ans: 143; 41·8 kc/s; 143° leading

(d) What would be the mid-frequency gain if the capacitor C_2 were removed?

Ans: 89

(e) What would be the gain at $\omega = 100$ if C_2, C_3 and C_4 were all too large to have any effect and if C_1 were equal to $0.01\ \mu F$? What would be the phase of v_a relative to v_g?

Ans: 101; 135° lagging

3. An amplifier, which is not required to deliver appreciable power, is to be constructed of two similar stages in tandem and the minimum ratio of output voltage to input voltage is to be 100. In the absence of negative feedback, each stage gives a voltage amplification lying in the range between 80 and 100 times, depending on the age of the valves. It may be assumed that output and input voltages are 180° out of phase with each other.

Negative feedback is to be applied to reduce the variation of total amplification during the lives of the valves and the following possible arrangements are considered:

(a) negative feedback is applied in a similar manner to each stage separately;
(b) negative feedback is applied from the output of the second stage to the input of the first stage.

What is the maximum variation in amplification that can occur in each of these cases?

Ans: (a) 100–105·3; (b) 100–100·6

4. Derive the steady state oscillation conditions for the circuit of Fig. 3, assuming that r is much smaller than R and is also much smaller than the reactance of C. The mutual conductance of the valve is g.

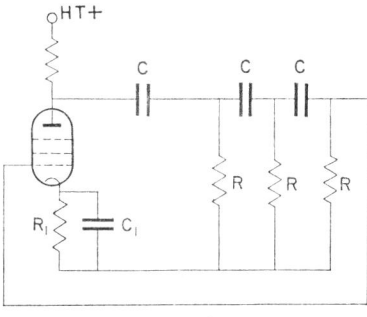

Fig. 3

Ans: $\omega = 1/\sqrt{(6)}RC$, $gr = 29$

5. In a test on an amplifier with negative feedback, the feedback loop is disconnected and separate measurements are made of the quantities A and β at a number of frequencies. At frequencies above 100 kc/s, the Nyquist diagram is found to lie to the right of the j-axis and the values of A and β, expressed in polar form, are as follows:

		100	150	200	300	400	500	800	5000 kc/s
A	Amplitude	10	8·8	8·5	8·2	7·9	6·0	1·7	0
	Phase angle	90°	65°	44°	19°	−3°	−21°	−48°	—
β	Amplitude	0·1	0·1	0·1	0·1	0·1	0·1	0·2	0·4
	Phase angle	0	0	2°	5°	9°	12°	20°	50°

Plot this portion of the Nyquist diagram and determine the range of frequencies over which the complete amplifier will be regenerative.

If the amplitude of A were increased by a constant factor k without changing the phase angle at any frequency, what would be the least value of k to cause spontaneous oscillation and at approximately what frequency would the oscillation occur?

Ans: From 150–5000 kc/s; 1·33; 440 kc/s

6. The current amplifier of Fig. 4 is supplied with 100 mA r.m.s. What is the current in R_L? What are the a.c., voltage and power gains?

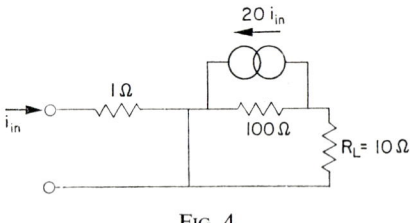

Fig. 4

Ans: 1·82 A r.m.s.; 18·2, 182, 3320

7. If the current source supplying the amplifier of Qu. 1 had a parallel resistance of 9, what would the alternating current in the load be, and what would be the overall current, voltage and power gains?

Ans: 1·62 A r.m.s.; 16·2, 182, 3320

8. Calculate the input and output impedance for the current amplifier of Fig. 5.

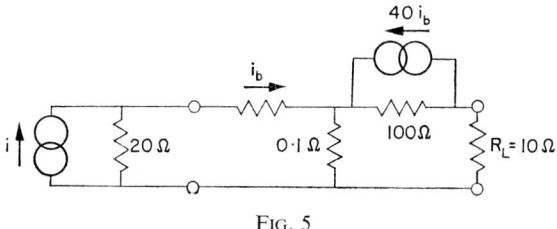

FIG. 5

Ans: 4·73 Ω, 118 Ω

9. Calculate the alternating current in R_L if, in Fig. 5, $i = 0.1$ A r.m.s.

Ans: 2·98 A r.m.s.

10. Two alternative models of a current amplifier are given in Fig. 6. Show that they are equivalent if $a = \beta/(\beta + 1)$ and $r_c = (1 + \beta)r_d$. (Hint: compare voltages across the dependent sources.)

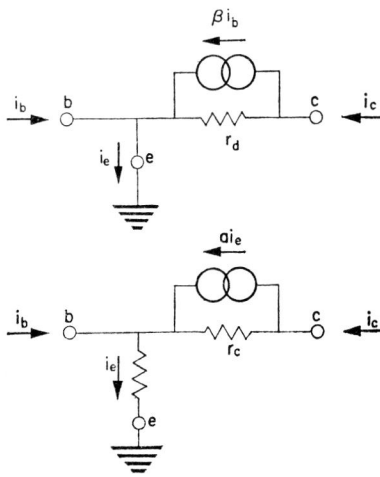

FIG. 6

11. Calculate the ratios

$$\frac{\text{a.c. load current}}{\text{input current}}, \quad \frac{\text{a.c. load voltage}}{\text{source voltage}}, \quad \frac{\text{a.c. load power}}{\text{a.c. input power}}$$

for the transistor circuit of Fig. 7.

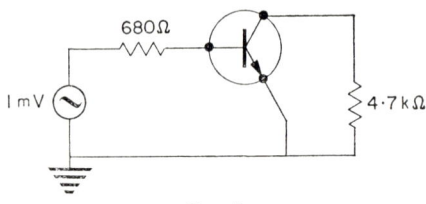

Fig. 7

Ans: 93·3, 175, 22,400

12. A certain voltage amplifier has input resistance = 100 kΩ, voltage gain = 20, and output resistance = 5 kΩ. It is desired to amplify 4 mV RMS signals coming from a source of internal resistance 5 kΩ and to develop at least 20 mW in headphones of resistance = 1 kΩ.

(a) How many stages of amplification are needed?

(b) What value of coupling capacitor is needed between the last amplifier stage and the headphones if the output can drop to 10 mW at 30 Hz?

(c) What length of screened cable can be used between the source and the first amplifier stage if the capacity between conductors in the cable is 60 pF/metre and if the output can be allowed to drop to 10 mW at 16 kHz?

Ans: (a) 3; (b) 1 μF; (c) 35 m.

13. Explain what is meant by *negative feedback* in an amplifier and discuss its effect on the performance of the amplifier.

Two identical voltage amplifiers have gains which may vary from 50 to 100. The gain is produced without phase shift.

If the amplifiers are to be used in series, with negative feedback of such magnitude that the overall gain does not fall below 100, determine the amount of feedback required and the maximum variation in gain (a) when feedback is applied overall, and (b) when equal feedbacks are applied to each amplifier separately.

Ans: (a) −0·0096, 100 to 103; (b) −0·08, 100 to 123.

14. The circuit of an a.c. amplifier is shown in Fig. 8(a). If the equivalent circuit used to represent the transistor is as shown in Fig. 3(b), and no current is taken at v', show that the collector and base currents are related by $i_c = i_b (\beta r_d - 1000)/(r_d + 6000)$.

Hence or otherwise obtain an expression for the open-circuit voltage gain, $-v'/v$, and find the lowest value for β which can be allowed if $r_d = 20000\,\Omega$, and the gain is to be greater than 4·8.

The components R and C do not influence the relevant a.c. conditions.

Ans: $5/\left(1 + \dfrac{26}{20\beta-1}\right)$, 31·25

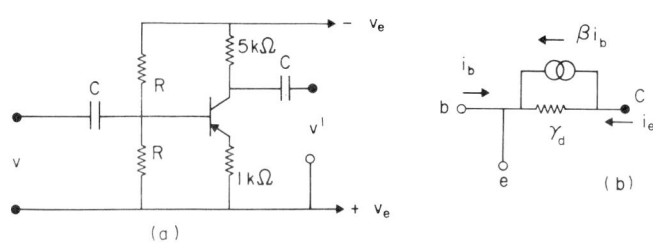

FIG. 8

15. In Fig. 9, the box represents an amplifier of gain -1000, input impedance 500 kΩ and negligible output impedance.

$$R_1 = 9\text{ k}\Omega, \; R_2 = 1\text{ k}\Omega$$

Calculate the voltage gain, and input impedance of the amplifier with feedback.

Ans: $\dfrac{-1000}{101}$, 50·5 MΩ

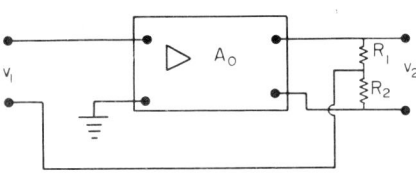

FIG. 9

16. For the amplifier of Fig. 9, the gain without feedback at a certain frequency is $A_o = 10^4\underline{/-150°}$. Given $R_1 + R_2 = 10^5\,\Omega$ and $R_2 = 10^2\,\Omega$, find the gain and phase angle of v_2/v_1.

State the assumptions that you make to solve the problem with the given data.

Ans: $920\underline{/-177°}$.

17. A two-stage f.e.t. oscillator uses the phase-shifting network shown in Fig. 2. If the input resistance of the amplifier is very high, prove

$$v_2/v_1 = 1/\{3 + j(\omega RC - 1/\omega RC)\}.$$

Show that the frequency of oscillation is $f = 1/2\pi RC$ and that the gain of the amplifier, A must exceed 3.

18. The oscillator shown in Fig. 10 is made with an amplifier of input resistance $= 1$ MΩ and output resistance $= 10$ kΩ and capacitors of $0\cdot01$ μF. What resistors are wanted for an oscillation frequency of 160 Hz?

Ans: 90 kΩ, 110 kΩ.

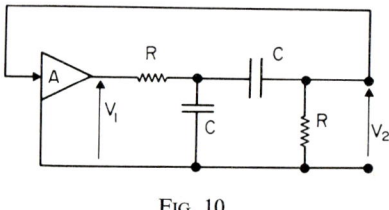

FIG. 10

19. Fig. 11 shows the circuit of a tuned anode oscillator. The tuned circuit has inductance $L = 150$ μH and capacitance $C = 200$ pF whilst the resistance R represents losses corresponding to a Q-factor of 50. The valve is operated at an anode slope resistance of 5 kΩ and an amplification factor of 30.

Determine the minimum value of the mutual inductance M in order that oscillations shall be sustained.

Ans: 5·2 μH.

Explain what happens if M exceeds this value.

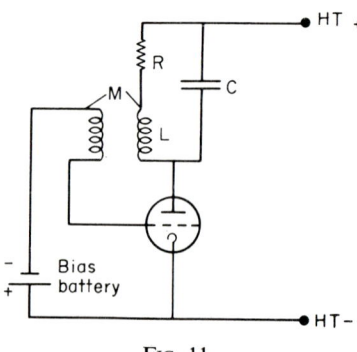

FIG. 11

Index

351